NONSMOOTH OPTIMIZATION

NONSMOOTH OPTIMIZATION

Analysis and Algorithms with Applications to Optimal Control

Marko M. Mäkelä & Pekka Neittaanmäki
Department of Mathematics
University of Jyväskylä
Finland

World Scientific
Singapore • New Jersey • London • Hong Kong

Published by

World Scientific Publishing Co. Pte. Ltd.

5 Toh Tuck Link, Singapore 596224

USA office: 27 Warren Street, Suite 401-402, Hackensack, NJ 07601

UK office: 57 Shelton Street, Covent Garden, London WC2H 9HE

Library of Congress Cataloging-in-Publication Data
Mäkelä, Marko M.
 Nonsmooth optimization: analysis and algorithms with applications to optimal control /
Marko M. Mäkelä and Pekka Neittaanmäki.
   p. cm.
 Includes bibliographical references and index.
 ISBN-13 978-981-02-0773-1 -- ISBN-10 981-02-0773-5
 ISBN-13 978-981-02-3690-8 (pbk) -- ISBN-10 981-02-3690-5 (pbk)
 1. Mathematical optimization. 2. Mathematical analysis.
 I. Neittaanmäki, P. (Pekka) II. Title.
 QA402.5.M315 1992
 519.3--dc20   91-48073
                                                                            CIP

British Library Cataloguing-in-Publication Data
A catalogue record for this book is available from the British Library.

Preface

In practical applications of optimization we often get into the situation where the objective function to be minimized/maximized is not necessarily differentiable. The source of the nonsmoothness may be the objective function itself, its possible interior function or both. For example, economics tax models typically consists of several different elements which at their intersections have discontinuous gradients. In optimal control problems governed by partial differential systems the smoothness may be caused by the technological constraints. We may add that there also exist so called stiff problems which are smooth analytically but nonsmooth numerically, where the gradient varies too rapidly and the classical methods will fail. Thus we cannot directly use the methods demanding differential information while the usual methods that do not require gradients are often very inefficient. Instead of gradients we must use the so called generalized gradients (or subgradients) which allow us to generalize the effective smooth derivative methods for nonsmooth problems.

The aim of this book is to introduce various methods for nonsmooth optimization and to apply these methods to solve discretized nonsmooth optimal control problems of systems governed by boundary value problems. The material of this work is divided into three parts, which are organized as follows. The first part consists of nonsmooth analysis and optimization theory. The material of this part is planned to be at the introductory level. We concentrate only on the results relevant for optimization and have simplified the presentation by considering only functions defined on finite dimensional Euclidean spaces. To give as self-contained a description as possible, we shall give relatively detailed proofs of all the theorems and lemmas.

In the second part we consider the methods of nonsmooth optimization. After a short survey of the methods and their development we shall construct a special method for our problem. After a detailed algorithm description we shall report on some numerical results attained by our implementation. We have tested some traditional examples from nonsmooth optimization literature and compared the results with other methods.

The third part is devoted to optimal control problems and their numerical realization by nonsmooth optimization methods. We first consider distributed parameter control problems for elliptic systems with state constraints. In the following chapter we study so called optimal shape design problems. In these optimal control problems the objective is to minimize or maximize certain criterion function involving the solution of the state system (partial differential system, for example) with respect to coefficients or the right hand side of the state system (distributed control problems), boundary values (boundary control) or with respect to the domain of definition of the state system (optimal shape design). After we have performed the discretization we are in a position to apply methods of nonsmooth optimization. In some examples of nonsmooth optimal control problems we have compared the performance of nonsmooth optimization methods with the results obtained by conventional methods (regularization by some appropriate smoothing technique like penalty methods and application of smooth optimization methods).

The reader interested only in nonsmooth optimization and their numerical study may confine himself to the first two parts. This book is a self-contained elementary study for nonsmooth analysis and optimization, and their use in the solution of nonsmooth optimal control problems.

The authors are indebted to J. Haslinger, K. Kiwiel, R. Mäkinen, T. Männikkö, A. Eljendy, T. Roubížek and F. Töltzsch for their valuable comments. We are also indebted to A. Roikonen, T. Räisänen and J. Toivanen for their help in preparing this book. We also acknowledge the Academy of Finland for their financial support.

Finally, we would like to express our thanks to World Scientific Publishing House for their friendly co-operation.

Jyväskylä, Finland   M.M. Mäkelä

January 1992   P. Neittaanmäki

List of Symbols

$\mathbf{R}^n$ — n-dimensional Euclidean space

$\mathbf{C}$ — set of complex numbers

$\mathbf{N}$ — set of natural numbers

$\mathbf{Z}$ $(\mathbf{Z}_+)$ — set of integers (set of non-negative integers)

(a, b) — open interval

$[a, b]$ — closed interval

cl G — closure of set G

int G — interior of set G

conv G — convex hull of set G

card G — cardinality of set G

$T_G(x)$ — tangent cone of set G at x

$N_G(x)$ — normal cone of set G at x

$\Pi[x, G]$ — projection of x to set G

$d_G(x)$ — $= \inf\{\|x - c\| \mid c \in G\}$, $x \in \mathbf{R}^n$, distance function

x^{T} — transposed vector

B^{T} — transposed matrix

$x^{\mathrm{T}}y$ — inner product of x and y

ln — natural logarithm

det B — determinant of matrix B

δ_{ij} — Kronecker's symbol:
$\delta_{ij} = 1$ for $i = j$, otherwise $\delta_{ij} = 0$

$B(x, r)$ — $\{y \in X \; : \; \|y - x\| < r\}$ open ball
with radius r and central point x

$S(x, r) = \partial B(x, r)$ — $\{y \in X \; : \; \|y - x\| = r\}$ boundary of $B(x, r)$

$\partial f / \partial x_i$ — partial derivative of f with respect to x_i

$\frac{\partial}{\partial n} f$ — normal derivative of f on the boundary

$D^\alpha f$ — α-th distributional derivative of f (α multi-index)

$f'(x; v)$ — directional derivative of f at x in the direction v

$f^0(x; v)$ — generalized directional derivative of f at x in
the direction v

∂f — subdifferential of f

$\partial_c f$ — subdifferential of convex f

vii

$\partial_\varepsilon f(x)$	ε-subdifferential of f
$\partial_\varepsilon^G f(x)$	Goldstein ε-subdifferential of f
∇f	gradient of function $f : \mathbf{R}^n \to \mathbf{R}$;
	$\nabla f = (\frac{\partial}{\partial x_1} f, ..., \frac{\partial}{\partial x_n} f)$
$\nabla f \; (= \nabla_x f(x))$	Jacobian of function $f : \mathbf{R}^n \to \mathbf{R}^n$;
	$\nabla f = (\frac{\partial}{\partial x_i} f_j(x))_{i,j=1}^n$
$\Delta f = \nabla^{\mathrm{T}} \nabla f$	Laplace operator
epi f	epigraph of f
lev f	level set of f
$\arg\min f(x)$	a point where f has its minimum value
X'	dual space of X
$x_k \rightharpoonup x$	weak convergence
$x_k \downarrow 0$	$x_k \to 0_+$
$\|x\| \; (= \|x\|_X)$	norm on Banach space X
$\|f\|_{X'} = \sup_{\|x\| < 1} |f(x)|$	norm in dual space X'
$\|f\| = \max_{1 \le i \le n} \max_{x \in X} |f_i(x)|$	supremum norm
$C(\Omega)$	$\{f : \Omega \to \mathbf{R} \mid f \text{ is continuous with the supremum norm}\}$
$C^m(\Omega)$	a space of functions $f : \Omega \to \mathbf{R}$ with
	continuous derivatives up to order m
$C_0^\infty(\Omega)$	a space of functions from $C^\infty(\Omega)$
	with compact support
$L^p(\Omega), \; p < \infty$	Lebesgue space of measurable functions
	f defined on Ω for which $[\int_\Omega |f|^p \; d\Omega]^{1/p}$ is finite
$H^m(\Omega)$	$\{f \in L^2(\Omega) \mid D^\alpha f \in L^2(\Omega), \; |\alpha| \le m\}$
$L^\infty(\Omega)$	Lebesgue space of measurable essentially
	bounded functions defined on Ω
$P_k(\Omega)$	space of polynomials of degree at most k defined on Ω
$\mathcal{L}(\mathbf{R}^n, \mathbf{R})$	set of linear mappings from $\mathbf{R}^n$ to $\mathbf{R}$
$\mathcal{P}(\mathbf{R}^n)$	set of all subsets in $\mathbf{R}^n$
$c, \hat{c}, c_i, \ldots$	generic constant
$\exists$	there exist(s)
$\forall$	for all
a.a.	almost all
a.e.	almost everywhere
s.t.	such that

Contents

Part II
Nonsmooth Optimization

Part III
Nonsmooth Optimal Control

Part I

Nonsmooth Analysis

Chapter 1

Introduction

The history of nonsmooth (not necessarily differentiable) analysis is quite short but very intense. The classical theory of optimization has always been connected to differentiation and strong regularity assumptions. However, these assumptions are often too demanding for practical applications, due to the "nonsmoothness" of natural phenomena. But necessity is the mother of invention: typically these nonsmooth problems are approximated by smooth ones and the developed theory is applied to them. This approach, however, causes other difficulties in the form of approximation errors, since the approximations may be too rough. This difficulty was one initiator of the theory of nonsmooth analysis.

Everyone who has dealt with optimization knows that convexity is a pleasant property and that optimization has a very clear geometrical interpretation. Thus it was natural that the theory was first developed for convex functions and the treatment was quite geometrical. In optimization theory the meaning of differentiation is to locally linearize the given differentiable function in the sense that the hyperplane generated by the gradient is the tangent plane of the graph of the function. For convex functions these linearizations are always lower approximations. We get a lower piecewise linear approximation by taking a maximum over the linearizations defined at several points and by taking the maximum over all points we get the original convex differentiable function. These ideas were generalized for nonsmooth convex functions by defining the concepts of *subgradient* and *subdifferential*. A subgradient at a fixed point is a vector which has the property that the hyperplane at that point generated by the vector is a lower approximation to the function; the set of all subgradients at that point is called subdifferential. These new concepts made it possible to obtain the same approximation properties as in the smooth case.

The next long step in nonsmooth analysis was the generalization of subgradients to nonconvex functions. In practical optimization a suitable and sufficiently general class of nonconvex functions are the locally Lipschitz continuous functions. In the convex case differentiation was thought of as the linearization of a function. Now, for nonconvex functions we can think of differentiation as convexification. The basic ideas are the same but the result is much more theoretical. As in the convex case, the links between differentiation and geometry are strong: instead of

3

subgradients and subdifferentials of the functions we can talk about tangents and normals of epigraph sets.

This first part is mainly based on the works of **Rockafellar** (1970, 1981) and **Clarke** (1983, 1989). Our aim is to present the theory of nonsmooth analysis for optimization in a compact and "easy-to-understand" form. The reader is assumed to have some basic knowledge of linear algebra, elementary real analysis and smooth nonlinear optimization. To give as self-contained a description as possible, we shall define every concept used and prove all theorems and lemmas. We try to explain how everything has taken place. For this reason we recall some basic results and notions of smooth analysis. Chapter 2 is devoted entirely to convex analysis. First we generalize the differential concepts for convex, not necessarily differentiable functions and after that we take a geometrical viewpoint for the problem. In section 2.3 we show the connection between differentiation and geometry. After this flashback it is easier to understand the theoretical definitions and the roots of the concepts in the next chapters.

In chapter 3 we generalize the convex differential theory to locally Lipschitz continuous functions. As in classical differential theory, direct computation of derivatives from the definition is something one seeks to avoid. For this reason we are going to generalize such familiar derivation rules as the mean value theorem and the chain rule for subdifferential theory. We shall also present a very useful representation of subdifferentials as limits of ordinary gradients. At the end of the chapter we shall define so called ε-subdifferentials, which approximate the ordinary subdifferentials, and generalized Jacobians, which in turn generalize the derivative of vector-valued functions.

In chapter 4 we generalize notions of the convex geometry for nonconvex sets. Tangents and normals are generalized in a natural way and several representations of them are presented. The same links between differentiation and geometry as in the convex case are proved to be true.

In chapter 5 we concentrate on the theory of nonsmooth optimization. We shall generalize the classical optimality condition for differentiable function: at the extremum points the gradient must be zero. We shall give the necessary conditions for a Lipschitz function to attain its local minimum both in the unconstrained and constrained cases. For convex functions these conditions are also sufficient and the minimum is global. Moreover we are going to linearize the unconstrained and constrained optimization problems by using the subgradient information.

1.1. Notations

All the vectors are considered as column vectors. We denote by $x^T y$ and $\|x\|$, respectively, the usual inner product and norm in n-dimensional real Euclidean space $\mathbf{R}^n$, i.e.

$$\|x\| = (x^T x)^{\frac{1}{2}} \qquad \text{for} \quad x, y \in \mathbf{R}^n. \tag{1.1}$$

The open ball with center x and radius $\lambda > 0$ is denoted by $B(x; \lambda)$. We denote by $[x, y]$ the closed line segment joining x and y, i.e.

$$[x, y] = \{z \in \mathbf{R}^n \mid z = \lambda x + (1 - \lambda)y \quad \text{for} \quad 0 \le \lambda \le 1\}, \tag{1.2}$$

and by (x, y) the corresponding open line segment. A set $C \subset \mathbf{R}^n$ is called *convex* if $[x, y] \subset C$ for all x and y belonging to C. A linear combination $\sum_{j=1}^k \lambda_j x^j$ is called a *convex combination* of points $x_1, \ldots, x_k$ in $\mathbf{R}^n$ if each $\lambda_j \ge 0$ and $\sum_{j=1}^k \lambda_j = 1$. The *convex hull* of a set $C \subset \mathbf{R}^n$, denoted by $\operatorname{conv} C$, is the set of all convex combinations of points in C. In other words, $\operatorname{conv} C$ is the smallest convex set containing C and C is convex if and only if C and $\operatorname{conv} C$ coincide. The closure of C is denoted by $\operatorname{cl} C$ and the set of all subsets in $\mathbf{R}^n$ (power set) is denoted by $\mathcal{P}(\mathbf{R}^n)$.

The set C is said to be a *cone* if it contains all positive multiples of its elements, i.e. $x \in C$ and $\lambda > 0$ also imply that $\lambda x \in C$.

A function $f : \mathbf{R}^n \to \mathbf{R}$ is said to be *convex* if

$$f(\lambda x + (1 - \lambda)y) \le \lambda f(x) + (1 - \lambda)f(y), \tag{1.3}$$

whenever x and y are in $\mathbf{R}^n$ and $\lambda \in [0, 1]$. If strict inequality holds for $x \ne y$ and $\lambda \in (0, 1)$, then f is called *strictly convex*. A function f is *locally Lipschitz with constant K* at $x \in \mathbf{R}^n$ if there exists some $\varepsilon > 0$ such that

$$|f(y) - f(z)| \le K\|y - z\| \qquad \text{for all} \quad y, z \in B(x; \varepsilon). \tag{1.4}$$

Furthermore, f is *positively homogeneous* if $f(\lambda x) = \lambda f(x)$ for all $\lambda \ge 0$ and *subadditive* if $f(x + y) \le f(x) + f(y)$ for all x and y in $\mathbf{R}^n$. Note that a positively homogeneous, subadditive function is always convex.

A function $f : \mathbf{R}^n \to \mathbf{R}$ is said to be *upper semicontinuous* at x if for every sequence (x_i) converging to x we have

$$\limsup_{i \to \infty} f(x_i) \le f(x) \tag{1.5}$$

and *lower semicontinuous* if

$$f(x) \leq \liminf_{i \to \infty} f(x_i). \tag{1.6}$$

Note that an upper and lower semicontinuous function is continuous.

Next we recall some basic results of smooth differential theory. The *directional derivative* of f at x in the direction $v \in \mathbf{R}^n$ is defined by

$$f'(x; v) = \lim_{t \downarrow 0} \frac{f(x + tv) - f(x)}{t}. \tag{1.7}$$

If f is differentiable at x, then the directional derivative, which exists in every direction $v \in \mathbf{R}^n$, is a linear function of v and we have the relation

$$f'(x; v) = \nabla f(x)^{\mathrm{T}} v, \tag{1.8}$$

where $\nabla f(x) \in \mathbf{R}^n$ is the *gradient vector* of f at x. Moreover, by Taylor's formula we have

$$\begin{aligned} f(x + d) &= f(x) + \nabla f(x)^{\mathrm{T}} d + \|d\| \varepsilon(d) \\ &= f(x) + f'(x; d) + \|d\| \varepsilon(d) \qquad \text{for all} \quad d \in \mathbf{R}^n, \end{aligned} \tag{1.9}$$

where $\varepsilon(d) \to 0$ as $\|d\| \to 0$. If, in addition, f is convex, then we have

$$f(y) \geq f(x) + \nabla f(x)^{\mathrm{T}} (y - x) \qquad \text{for all} \quad y \in \mathbf{R}^n. \tag{1.10}$$

We denote by Ω_f the set in $\mathbf{R}^n$ where f is not differentiable.

1.2. Basic Results

We collect below some classical results from functional and real analysis.

Theorem 1.2.1 (Hahn-Banach). *Let X be a real linear vector space and $Y \subset X$ be a linear subspace. If $p : X \to \mathbf{R}$ is a positively homogeneous, subadditive functional and $f : Y \to \mathbf{R}$ a linear functional such that $f(y) \leq p(y)$ for all $y \in Y$, then there exists a linear functional $F : X \to \mathbf{R}$ such that*

$$F(y) = f(y) \quad \text{for all} \quad y \in Y \qquad \text{and} \qquad F(x) \leq p(x) \quad \text{for all} \quad x \in X. \tag{1.11}$$

Proof. See **Friedman** (1982) p. 151. □

Theorem 1.2.2. *The convex hull of a compact set is compact.*

Proof. See **Roberts and Varberg** (1973) p. 78. □

Theorem 1.2.3. *Let C and D be compact, convex sets in $\mathbf{R}^n$. Then $C \subset D$ if and only if*

$$\max_{c \in C} c^T x \leq \max_{d \in D} d^T x \qquad \text{for all} \quad x \in \mathbf{R}^n. \tag{1.12}$$

Proof. See **Rockafellar** (1970) p. 113. □

Theorem 1.2.4. *Let U be an open set in $\mathbf{R}^n$ and $f : U \to \mathbf{R}$. If*

$$\limsup_{t \downarrow 0} \left| \frac{f(x + tv) - f(x)}{t} \right| < \infty \qquad \text{for all} \quad x \in U \quad \text{and} \quad v \in \mathbf{R}^n, \tag{1.13}$$

then f is differentiable almost everywhere on U.

Proof. See **Kantorovich** (1947) pp. 233–236. □

Theorem 1.2.5 (Bolzano–Weierstrass). *If a bounded set C in $\mathbf{R}^n$ contains infinitely many points, then there is at least one point in $\mathbf{R}^n$ which is a cluster point of C.*

Proof. See **Apostol** (1974) p. 54. □

Theorem 1.2.6 (Fubini). *If f is a Lebesgue-measurable function on the Cartesian product $X \times Y$ of two finite dimensional real vector spaces and if either $\int \int |f| \, d\mu \, d\nu < \infty$ or $\int \int |f| \, d\nu \, d\mu < \infty$, then f is integrable on $X \times Y$ and*

$$\int \int f \, d\mu \, d\nu = \int \int f \, d\nu \, d\mu. \tag{1.14}$$

Proof. See **Friedman** (1982) p. 86. □

Chapter 2

Convex Analysis

The theory of nonsmooth analysis is based on convex analysis. For this reason we shall survey basic concepts and results of convexity from **Rockafellar** (1970) (for further readings see also **Roberts and Varberg** (1973)).

First we shall define subgradients and subdifferentials of convex functions and present some basic results. We shall take a geometrical viewpoint by examining tangents and normals of convex sets. To the end of this chapter we link these analytic and geometric concepts together.

2.1. Functions, Derivatives and Subgradients

In this section we shall generalize the classical derivative to convex but not necessarily differentiable functions. We start by giving an equivalent definition of a convex function.

Theorem 2.1.1 (Jensen's inequality). *A function* $f : \mathbf{R}^n \to \mathbf{R}$ *is convex if and only if*

$$f\left(\sum_{i=1}^{m} \lambda_i x_i\right) \leq \sum_{i=1}^{m} \lambda_i f(x_i), \tag{2.1}$$

whenever $x_i \in \mathbf{R}^n$, $\lambda_i \in [0,1]$ *for all* $i = 1, \ldots, m$ *and* $\sum_{i=1}^{m} \lambda_i = 1$.

Proof. Follows by induction from the definition of convex function. $\qquad\square$

Next we show that a convex function is locally Lipschitz.

Theorem 2.1.2. *Let* $f : \mathbf{R}^n \to \mathbf{R}$ *be a convex function. Then for any* x *in* $\mathbf{R}^n$, f *is locally Lipschitz at* x.

Proof. Let $u \in \mathbf{R}^n$ be arbitrary. We begin by proving that f is bounded on a neighborhood of u. Let $\varepsilon > 0$ and define the hypercube

$$S := \{y \in \mathbf{R}^n \mid |y_i - u_i| \leq \varepsilon \quad \text{for all} \quad i = 1, \ldots, n\}.$$

Let $u^1, \ldots, u^m$ denote the $m = 2^n$ vertices of S and let

$$M := \max\left\{f(u^i) \mid i = 1, \ldots, m\right\}.$$

8

Since each $y \in S$ can be expressed as $y = \sum_{i=1}^{m} \lambda_i u^i$ with $\lambda_i \geq 0$ and $\sum_{i=1}^{m} \lambda_i = 1$, by Theorem 2.1.1 we obtain

$$f(y) = f\left(\sum_{i=1}^{m} \lambda_i u^i\right) \leq \sum_{i=1}^{m} \lambda_i f(u^i) \leq M \sum_{i=1}^{m} \lambda_i = M.$$

Since $B(u; \varepsilon) \subset S$, we have an upper bound M of f on an ε-neighborhood of u, that is

$$f(x') \leq M \quad \text{for all} \quad x' \in B(u; \varepsilon).$$

Now let $x \in \mathbf{R}^n$, choose $\rho > 1$ and $y \in \mathbf{R}^n$ so that $y = \rho x$. Define

$$\lambda := 1/\rho \quad \text{and}$$
$$V := \{v \mid v = (1 - \lambda)(x' - u) + x, \quad \text{where} \quad x' \in B(u; \varepsilon)\}.$$

The set V is a neighborhood of $x = \lambda y$ with radius $(1 - \lambda)\varepsilon$. By convexity one has for all $v \in V$

$$f(v) = f((1 - \lambda)(x' - u) + \lambda y) = f((1 - \lambda)x' + \lambda(y + u - \tfrac{1}{\lambda}u))$$
$$\leq (1 - \lambda)f(x') + \lambda f(y + u - \tfrac{1}{\lambda}u).$$

Now $f(x') \leq M$ and $f(y + u - \tfrac{1}{\lambda}u) = \text{constant} =: K$ and so

$$f(v) \leq M + \lambda K.$$

That is, f is bounded above on a neighborhood of x.

Let us next show that f is also bounded below. Let $z \in B(x; (1 - \lambda)\varepsilon)$ and define $z' := 2x - z$. Then

$$\|z' - x\| = \|x - z\| \leq (1 - \lambda)\varepsilon.$$

Thus $z' \in B(x; (1 - \lambda)\varepsilon)$ and $x = (z + z')/2$. The convexity of f implies that

$$f(x) = f((z + z')/2) \leq \tfrac{1}{2}f(z) + \tfrac{1}{2}f(z'),$$

and

$$f(z) \geq 2f(x) - f(z') \geq 2f(x) - M - \lambda K$$

so that f is also bounded below on a neighborhood of x. Thus we have proved that f is bounded on a neighborhood of x.

Let $N > 0$ be a bound of $|f|$ so that

$$|f(x')| \le N \quad \text{for all} \quad x' \in B(x; 2\delta),$$

where $\delta > 0$, and let $x_1, x_2 \in B(x; \delta)$ with $x_1 \ne x_2$. Define

$$x_3 := x_2 + (\delta/\alpha)(x_2 - x_1),$$

where $\alpha := \|x_2 - x_1\|$. Then

$$\|x_3 - x\| = \|x_2 + (\delta/\alpha)(x_2 - x_1) - x\| \le \|x_2 - x\| + (\delta/\alpha)\|x_2 - x_1\|$$
$$< \delta + \frac{\delta}{\|x_2 - x_1\|}\|x_2 - x_1\| = 2\delta,$$

so $x_3 \in B(x; 2\delta)$. Solving for x_2 gives

$$x_2 = \frac{\delta}{\alpha + \delta}x_1 + \frac{\alpha}{\alpha + \delta}x_3,$$

and so by convexity

$$f(x_2) \le \frac{\delta}{\alpha + \delta}f(x_1) + \frac{\alpha}{\alpha + \delta}f(x_3).$$

Then

$$f(x_2) - f(x_1) \le \frac{\alpha}{\alpha + \delta}[f(x_3) - f(x_1)] \le \frac{\alpha}{\delta}|f(x_3) - f(x_1)|$$
$$\le \frac{\alpha}{\delta}(|f(x_3)| + |f(x_1)|).$$

Since $x_1, x_3 \in B(x; 2\delta)$ we have $|f(x_3)| < N$ and $|f(x_1)| < N$, so

$$f(x_2) - f(x_1) \le \frac{2N}{\delta}\|x_2 - x_1\|.$$

By changing the roles of x_1 and x_2 we have

$$|f(x_2) - f(x_1)| \le \frac{2N}{\delta}\|x_2 - x_1\|,$$

showing that the function f is locally Lipschitz at x.   □

Theorem 2.1.3. *Let* $f : \mathbf{R}^n \to \mathbf{R}$ *be a convex function with a Lipschitz constant* K *at* $x \in \mathbf{R}^n$. *Then*

(i) *the directional derivative in each direction $v \in \mathbf{R}^n$ exists and satisfies*

$$f'(x; v) = \inf_{t>0} \frac{f(x + tv) - f(x)}{t}, \tag{2.2}$$

(ii) *the function $v \mapsto f'(x; v)$ is positively homogeneous and subadditive on $\mathbf{R}^n$ with*

$$|f'(x; v)| \leq K\|v\|,$$

(iii) $f'(x; v)$ *is upper semicontinuous as a function of $(x; v)$ and Lipschitz with constant K as a function of v on $\mathbf{R}^n$,*

(iv) $-f'(x; -v) \leq f'(x; v)$.

Proof. (i) Let $v \in \mathbf{R}^n$ be an arbitrary direction. Define $\varphi : \mathbf{R} \to \mathbf{R}$ by

$$\varphi(t) := \frac{f(x + tv) - f(x)}{t}.$$

We begin by proving that φ is bounded below at t when $t \downarrow 0$. Let $\varepsilon > 0$ and let constants t_1 and t_2 be such that $0 < t_1 < t_2 < \varepsilon$. By the convexity of f one has

$$
\begin{aligned}
\varphi(t_2) - \varphi(t_1) &= \frac{1}{t_1 t_2}[t_1 f(x + t_2 v) - t_2 f(x + t_1 v) + (t_2 - t_1)f(x)] \\
&= \frac{1}{t_1}\left\{ \left(\frac{t_1}{t_2} f(x + t_2 v) + (1 - \frac{t_1}{t_2})f(x) \right) \right. \\
&\quad \left. - f\left(\frac{t_1}{t_2}(x + t_2 v) + (1 - \frac{t_1}{t_2})x \right) \right\} \\
&\geq 0,
\end{aligned}
$$

thus the function $\varphi(t)$ decreases as $t \downarrow 0$. Then for all $0 < t < \varepsilon$ one has

$$
\begin{aligned}
\varphi(t) - \varphi(-\varepsilon/2) &= \frac{\frac{1}{2}f(x + tv) + \frac{1}{2}f(x) + \frac{t}{\varepsilon}f(x - \frac{\varepsilon}{2}v) + (1 - \frac{t}{\varepsilon})f(x) - 2f(x)}{t/2} \\
&\geq \frac{\frac{1}{2}f(x + \frac{t}{2}v) + \frac{1}{2}f(x - \frac{t}{2}v) - f(x)}{t/4} \\
&\geq \frac{f(x) - f(x)}{t/4} = 0,
\end{aligned}
$$

which means that the function φ is bounded below for $0 < t < \varepsilon$. This implies that there exists the limit

$$\lim_{t \downarrow 0} \varphi(t) = f'(x; v) \quad \text{for all} \quad v \in \mathbf{R}^n$$

and since the function $\varphi(t)$ decreases as $t \downarrow 0$ we deduce that

$$f'(x; v) = \inf_{t>0} \frac{f(x + tv) - f(x)}{t}.$$

(ii) We start by proving the inequality in (ii). From the Lipschitz condition we obtain

$$|f'(x; v)| \leq \lim_{t \downarrow 0} \frac{|f(x + tv) - f(x)|}{t}$$

$$\leq \lim_{t \downarrow 0} \frac{K\|x + tv - x\|}{t}$$

$$\leq K\|v\|.$$

Next we show that $f'(x; \cdot)$ is positively homogeneous. To see this, let $\lambda > 0$. Then

$$f'(x; \lambda v) = \lim_{t \downarrow 0} \frac{f(x + t\lambda v) - f(x)}{t}$$

$$= \lim_{t \downarrow 0} \lambda \cdot \left\{ \frac{f(x + t\lambda v) - f(x)}{t\lambda} \right\}$$

$$= \lambda \cdot \lim_{t \downarrow 0} \frac{f(x + t\lambda v) - f(x)}{t\lambda} = \lambda \cdot f'(x; v).$$

We turn now to the subadditivity. Let $v, w \in \mathbf{R}^n$ be arbitrary, then by convexity

$$f'(x; v + w) = \lim_{t \downarrow 0} \frac{f(x + t(v + w)) - f(x)}{t}$$

$$= \lim_{t \downarrow 0} \frac{f(\frac{1}{2}(x + 2tv) + \frac{1}{2}(x + 2tw)) - f(x)}{t}$$

$$\leq \lim_{t \downarrow 0} \frac{f(x + 2tv) - f(x)}{2t} + \lim_{t \downarrow 0} \frac{f(x + 2tw) - f(x)}{2t}$$

$$= f'(x; v) + f'(x; w).$$

Thus $v \mapsto f'(x; v)$ is subadditive, which establishes (ii).

(iii) Let (x_i), $(v_i) \subset \mathbf{R}^n$ be sequences such that $x_i \to x$, $v_i \to v$ and let K be the Lipschitz constant at x. Define the sequence $(t_i) \subset \mathbf{R}$ by $t_i := K\|x_i - x\|^{\frac{1}{2}} + \frac{1}{i}$, so that $t_i \to 0$ and $t_i > 0$. Now we have

$$f'(x_i; v_i) = \inf_{t>0} \frac{f(x_i + tv_i) - f(x_i)}{t} \leq \frac{f(x_i + t_i v_i) - f(x_i)}{t_i}$$

$$= \frac{f(x + t_i v) - f(x)}{t_i} + \frac{f(x_i + t_i v_i) - f(x + t_i v)}{t_i}$$

$$+ \frac{f(x) - f(x_i)}{t_i}$$

and as $i \to \infty$, we have by the Lipschitz condition that

$$\frac{|f(x_i + t_i v_i) - f(x + t_i v)|}{t_i} \leq \frac{K\|x_i - x\| + K\|t_i v_i - t_i v\|}{t_i}$$

$$\leq \|x_i - x\|^{\frac{1}{2}} + K\|v_i - v\| \longrightarrow 0$$

and

$$\frac{|f(x) - f(x_i)|}{t_i} \leq \frac{K\|x - x_i\|}{t_i} \leq \|x - x_i\|^{\frac{1}{2}} \longrightarrow 0.$$

On taking upper limits (as $i \to \infty$), we obtain

$$\limsup_{i \to \infty} f'(x_i; v_i) \leq \limsup_{i \to \infty} \frac{f(x + t_i v) - f(x)}{t_i} = f'(x; v),$$

which establishes the upper semicontinuity. To show the Lipschitz condition, let v and w in $\mathbf{R}^n$ be given. If $x + tv$ and $x + tw$ are in $B(x; \varepsilon)$, then

$$f(x + tv) - f(x + tw) \leq Kt\|v - w\|,$$

and thus

$$\lim_{t \downarrow 0} \frac{f(x + tv) - f(x)}{t} \leq \lim_{t \downarrow 0} \frac{f(x + tw) - f(x)}{t} + K\|v - w\|,$$

whence

$$f'(x; v) - f'(x; w) \leq K\|v - w\|.$$

Reversing the roles of v and w we get

$$f'(x; w) - f'(x; v) \leq K\|v - w\|.$$

Thus

$$|f'(x; v) - f'(x; w)| \leq K\|v - w\|.$$

(iv) By assertion (ii) we have

$$\tfrac{1}{2} f'(x; v) + \tfrac{1}{2} f'(x; -v) \geq f'(x; \tfrac{1}{2} v - \tfrac{1}{2} v) = 0,$$

which implies that $-f'(x; -v) \leq f'(x; v)$. $\qquad\qquad\square$

Next we define the subgradient and the subdifferential of a convex function. Note the analogy to the smooth differential theory mentioned in the introduction. Figure 2.1 illustrates the meaning of the definition of the subdifferential.

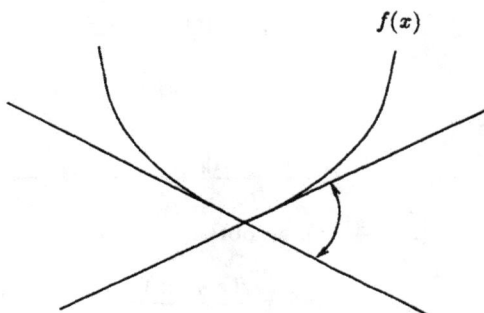

Figure 2.1.

Definition 2.1.4. The *subdifferential of a convex function* $f : \mathbf{R}^n \to \mathbf{R}$ at $x \in \mathbf{R}^n$ is the set

$$\partial_c f(x) = \{\xi \in \mathbf{R}^n \mid f(x') \geq f(x) + \xi^T(x' - x) \text{ for all } x' \in \mathbf{R}^n\}. \qquad (2.3)$$

Each element $\xi \in \partial_c f(x)$ is called a *subgradient* of f at x.

In the following we present the relationship between subdifferential and the ordinary directional derivative. As we shall see, it is enough to know either of the concepts; the other can be computed from that.

Theorem 2.1.5. Let $f : \mathbf{R}^n \to \mathbf{R}$ be convex. Then at every x we have

(i) $f'(x; v) = \max\{\xi^T v \mid \xi \in \partial_c f(x)\}$ for all $v \in \mathbf{R}^n$,

(ii) $\partial_c f(x) = \{\xi \in \mathbf{R}^n \mid f'(x; v) \geq \xi^T v \text{ for all } v \in \mathbf{R}^n\}$,

(iii) $\partial_c f(x)$ is a nonempty, convex and compact set such that $\partial_c f(x) \subset B(0; K)$, where K is the Lipschitz constant of f at x,

(iv) the point-to-set mapping $\partial_c f(\cdot) : \mathbf{R}^n \to \mathcal{P}(\mathbf{R}^n)$ is upper semicontinuous, i.e., if $y_i \to x$ and $\xi_i \in \partial_c f(y_i)$ for each i, then each accumulation point ξ of (ξ_i) is in $\partial_c f(x)$.

Proof. (i) By definition of the subdifferential, we deduce that for each $v \in \mathbf{R}^n$

$$\varphi(t) = \frac{f(x + tv) - f(x)}{t} \geq \frac{\xi^T tv}{t} = \xi^T v \quad \text{for all} \quad \xi \in \partial_c f(x),$$

so

$$f'(x;v) \geq \max\{\xi^T v \mid \xi \in \partial_c f(x)\}.$$

On the other hand, if there were $v_1 \in \mathbf{R}^n$ for which $f'(x;v_1) > \max\{\xi^T v_1 \mid \xi \in \partial_c f(x)\}$ there would be, by the Hahn-Banach Theorem (Theorem 1.2.1), some $\xi_1 \in \mathbf{R}^n$ such that for all $v \in \mathbf{R}^n$

$$f'(x;v) \geq \xi_1^T v \quad \text{and} \quad f'(x;v_1) = \xi_1^T v_1.$$

For $x' \in \mathbf{R}^n$ there exist $v \in \mathbf{R}^n$ and $t > 0$ such that $x' = x + tv$. It follows that

$$f(x') - f(x) = t\varphi(t) \geq tf'(x;v) \geq t\xi_1^T v = \xi_1^T tv = \xi_1^T(x - x') \quad \forall\, x' \in \mathbf{R}^n,$$

which implies that $\xi_1 \in \partial_c f(x)$ and

$$f'(x;v_1) > \max\{\xi^T v_1 \mid \xi \in \partial_c f(x)\} \geq \xi_1^T v_1 = f'(x;v_1),$$

which is impossible. Thus the first statement is true.

(ii) Set $K := \{\xi \in \mathbf{R}^n \mid f'(x;v) \geq \xi^T v \ \forall v \in \mathbf{R}^n\}$ and let $\xi \in K$ be arbitrary. Then it follows from convexity that, for all $y \in \mathbf{R}^n$, one has

$$\begin{aligned}
\xi^T y &\leq f'(x;y) \\
&= \lim_{t\downarrow 0} \frac{f((1-t)x + t(x+y)) - f(x)}{t} \\
&\leq \lim_{t\downarrow 0} \frac{(1-t)f(x) + tf(x+y) - f(x)}{t} \\
&= f(x+y) - f(x),
\end{aligned}$$

whenever $t \leq 1$. By choosing $y := x' - x$ we derive $\xi \in \partial_c f(x)$. On the other hand, if $\xi \in \partial_c f(x)$ then for all $y \in \mathbf{R}^n$

$$f'(x;y) = \lim_{t\downarrow 0} \frac{f(x+ty) - f(x)}{t} \geq \lim_{t\downarrow 0} \frac{\xi^T(x+ty-x)}{t} \geq \xi^T y.$$

Thus $\xi \in K$, which establishes (ii).

(iii) In view of Theorem 2.1.3(ii) the function $f'(x;\cdot) : \mathbf{R}^n \to \mathbf{R}$ is positively homogeneous and subadditive. Then by the Hahn-Banach Theorem (Theorem 1.2.1) there exists a vector $\xi \in \mathbf{R}^n$ such that $\xi^T v \leq f'(x;v)$ for all $v \in \mathbf{R}^n$, which means by assertion (ii) that $\partial_c f(x)$ is nonempty. To see the convexity let $\xi_1, \xi_2 \in \partial_c f(x)$ and $\lambda \in [0,1]$. Then by (i)

$$
\begin{aligned}
(\lambda \xi_1 + (1-\lambda)\xi_2)^T v &= \lambda \xi_1^T v + (1-\lambda)\xi_2^T v \\
&\leq \lambda f'(x;v) + (1-\lambda)f'(x;v) \\
&\leq f'(x;v),
\end{aligned}
$$

whence by (ii) $\lambda \xi_1 + (1-\lambda)\xi_2 \in \partial_c f(x)$ and so the subdifferential is convex. By assertion (ii) and Theorem 2.1.3(ii) we may calculate for an arbitrary $\xi \in \partial_c f(x)$ that

$$
\|\xi\|^2 = |\xi^T \xi| \leq |f'(x;\xi)| \leq K\|\xi\|,
$$

which means that $\partial_c f(x)$ is bounded and so for the compactness it suffices to show that it is also closed. To see this, let $(\xi_i) \subset \partial_c f(x)$ be a sequence such that $\xi_i \to \xi$. Then

$$
\xi^T v = (\lim_{i\to\infty} \xi_i)^T v = \lim_{i\to\infty} \xi_i^T v \leq \lim_{i\to\infty} f'(x;v) = f'(x;v),
$$

which shows that $\xi \in \partial_c f(x)$ and so $\partial_c f(x)$ is closed. This establishes (iii).

(iv) Finally, let $(y_i) \subset \mathbf{R}^n$ and $(\xi_i) \subset \partial_c f(y_i)$ for each i be given sequences such that $y_i \to x$ and $\xi_i \to \xi$. Then for all $v \in \mathbf{R}^n$ we have

$$
\xi^T v = (\lim_{i\to\infty} \xi_i)^T v = \lim_{i\to\infty} \xi_i^T v \leq \limsup_{i\to\infty} f'(y_i;v).
$$

By Theorem 2.1.3(iii) the function $f'(x; \cdot)$ is upper semicontinuous and

$$
\xi^T v \leq f'(x;v).
$$

Thus $\partial_c f(\cdot)$ is upper semicontinuous as a set valued mapping and the proof is complete. $\square$

The next theorem shows that the subdifferential really is a generalization of the classical derivative.

Theorem 2.1.6. *If $f : \mathbf{R}^n \to \mathbf{R}$ is convex and differentiable at $x \in \mathbf{R}^n$, then*

$$
\partial_c f(x) = \{\nabla f(x)\}. \tag{2.4}
$$

Proof. From the definition of differentiability we have

$$
f'(x;v) = \nabla f(x)^T v \quad \text{for all} \quad v \in \mathbf{R}^n,
$$

which implies, by Theorem 2.1.5, that $\partial_c f(x) = \{\nabla f(x)\}$. $\square$

We are now ready to present the main result of this section. It gives a representation to a convex function by using subgradients.

Theorem 2.1.7. *If $f : \mathbf{R}^n \to \mathbf{R}$ is convex then for all $y \in \mathbf{R}^n$*

$$f(y) = \max \{f(x) + \xi^{\mathrm{T}}(y - x) \mid x \in \mathbf{R}^n, \ \xi \in \partial_c f(x)\}. \tag{2.5}$$

Proof. Suppose that $y \in \mathbf{R}^n$ is an arbitrary point and $\eta \in \partial f(y)$. Let

$$S = \{f(x) + \xi_x^{\mathrm{T}}(y - x) \mid \xi_x \in \partial_c f(x), \ x \in \mathbf{R}^n\}.$$

By the definition of subdifferential of a convex function

$$f(y) \geq f(x) + \xi_x^{\mathrm{T}}(y - x) \quad \text{for all } x \in \mathbf{R}^n \text{ and } \xi_x \in \partial_c f(x),$$

which means that the set S is bounded from above and

$$\sup S \leq f(y).$$

On the other hand,

$$f(y) = f(y) + \eta^{\mathrm{T}}(y - y) \in S,$$

so $f(y) \leq \sup S$. Thus

$$f(y) = \max\{f(x) + \xi_x^{\mathrm{T}}(y - x) \mid \xi_x \in \partial_c f(x), \ x \in \mathbf{R}^n\}.$$

$\square$

2.2. Sets, Tangents and Normals

We start this section by recalling the definitions of a convex set and of the tangent cone to a convex set. In what follows we shall use the notation C for a convex subset of $\mathbf{R}^n$.

Definition 2.2.1. Let C be a nonempty subset of $\mathbf{R}^n$. The set C is said to be *convex* if

$$(1 - \lambda)x + \lambda y \in C, \tag{2.6}$$

whenever $x \in C$, $y \in C$ and $\lambda \in [0, 1]$. *The tangent cone of the convex set C at* $x \in C$ is given by the formula

$$T_C(x) := \{y \in \mathbf{R}^n \mid \text{there exist } t_i \downarrow 0 \text{ and } y_i \to y \text{ with } x + t_i y_i \in C\}. \tag{2.7}$$

The next theorem shows that the space of convex sets has some linear properties as a subspace of the power set $\mathcal{P}(\mathbf{R}^n)$ consisting of all subsets of $\mathbf{R}^n$.

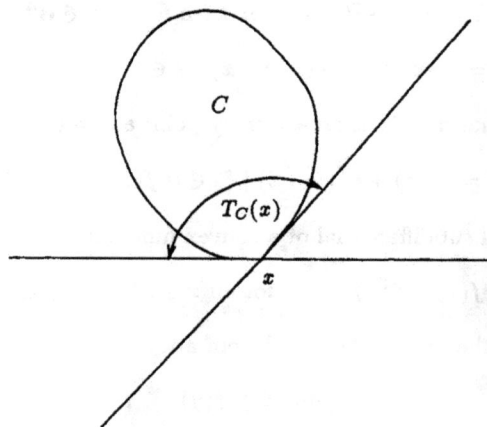

Figure 2.2. The tangent cone of convex set

Theorem 2.2.2. *Let C_1 and $C_2 \subset \mathbf{R}^n$ be convex subsets and $\mu_1, \mu_2 \in \mathbf{R}$. Then the set $\mu_1 C_1 + \mu_2 C_2$ is also convex.*

Proof. Let the points $x, y \in \mu_1 C_1 + \mu_2 C_2$ and $\lambda \in [0, 1]$. Then x and y can be written

$$\begin{cases} x = \mu_1 x_1 + \mu_2 x_2, & \text{where } x_1 \in C_1 \text{ and } x_2 \in C_2 \\ y = \mu_1 y_1 + \mu_2 y_2, & \text{where } y_1 \in C_1 \text{ and } y_2 \in C_2 \end{cases}$$

and

$$\lambda x + (1 - \lambda)y = \mu_1(\lambda x^1 + (1 - \lambda)y^1) + \mu_2(\lambda x^2 + (1 - \lambda)y^2)$$
$$\in \mu_1 C_1 + \mu_2 C_2.$$

Thus the set $\mu_1 C_1 + \mu_2 C_2$ is convex. $\qquad\square$

Theorem 2.2.3. *Let $C_i \subset \mathbf{R}^n$ be convex subsets for $i = 1, \ldots, m$. Then their intersection*

$$\bigcap_{i=1}^{m} C_i \quad \text{is also convex.} \tag{2.8}$$

Proof. Let $x, y \in \bigcap_{i=1}^{m} C_i$ and $\lambda \in [0, 1]$ be arbitrary. Because $x, y \in C_i$ and C_i is convex for all $i = 1, \ldots, m$, we have $\lambda x + (1 - \lambda)y \in C_i$ for all $i = 1, \ldots, m$. This implies that

$$\lambda x + (1 - \lambda)y \in \bigcap_{i=1}^{m} C_i$$

and the proof is complete.   □

Several elementary facts about the tangent cone will now be listed.

Theorem 2.2.4. *The tangent cone $T_C(x)$ of the convex set C is a closed convex cone containing zero.*

Proof. We begin by proving that $T_C(x)$ is closed. To see this, let (y^i) be a sequence in $T_C(x)$ which converges to $y \in \mathbf{R}^n$. We must show that $y \in T_C(x)$. The fact that $y^i \to y$ implies that for all $\varepsilon > 0$ there exists $i_0 \in \mathbf{N}$ such that

$$\|y - y^i\| < \varepsilon/2 \qquad \text{for all} \quad i \geq i_0.$$

On the other hand, $y^i \in T_C(x)$, so for each $i \in \mathbf{N}$ there exist sequence $(y^i_j) \subset \mathbf{R}^n$ and $(t^i_j) \subset \mathbf{R}$ so that $y^i_j \to y^i$, $t^i_j \downarrow 0$ and $x + t^i_j y^i_j \in C$ for all $j \in \mathbf{N}$. Then there exist $j^i_y \in \mathbf{N}$ and $j^i_t \in \mathbf{N}$ such that for all $i \in \mathbf{N}$

$$\|y^i - y^i_j\| < \varepsilon/2 \qquad \text{for all} \quad j \geq j^i_y$$

and

$$|t^i_j| < 1/i \qquad \text{for all} \quad j \geq j^i_t.$$

Now we choose $j^i := \max\{j^i_y, j^i_t\}$. Then $t^i_{j^i} \downarrow 0$ and for all $i \geq i_0$

$$\|y - y^i_{j^i}\| \leq \|y - y^i\| + \|y^i - y^i_{j^i}\| \leq \varepsilon/2 + \varepsilon/2 = \varepsilon,$$

which implies that $y^i_{j^i} \to y$ and, moreover, $x + t^i_{j^i} y^i_{j^i} \in C$. By the definition of the tangent cone, this means that $y \in T_C(x)$ and so $T_C(x)$ is closed.

Evidently $0 \in T_C(x)$, so we continue by proving that $T_C(x)$ is a cone. If $y \in T_C(x)$ is arbitrary then there exist sequences $(y_j) \subset \mathbf{R}^n$ and $(t_j) \subset \mathbf{R}$ such that $y_j \to y$, $t_j \downarrow 0$ and $x + t_j y_j \in C$ for all $j \in \mathbf{N}$. Let $\lambda > 0$ be fixed and define $y'_j := \lambda y_j$ and $t'_j := t_j/\lambda$. Since $t'_j \downarrow 0$,

$$\|y'_j - \lambda y\| = \lambda \|y_j - y\| \longrightarrow 0 \quad \text{whenever} \quad j \to \infty$$

and

$$x + t'_j y'_j = x + \frac{t_j}{\lambda} \cdot \lambda y_j \in C$$

it follows that

$$\lambda y \in \{y \in \mathbf{R}^n \mid \text{there exists } t_j \downarrow 0 \text{ and } y_j \to y \text{ with } x + t_j y_j \in C\} = T_C(x).$$

Thus $T_C(x)$ is a cone.

For convexity let $\lambda \in [0,1]$ and $y^1, y^2 \in T_C(x)$. We need to show that $y :=
(1-\lambda)y^1 + \lambda y^2$ belongs to $T_C(x)$. By the definition of $T_C(x)$ there exist sequences
$(y_j^1), (y_j^2) \subset \mathbf{R}^n$ and $(t_j^1), (t_j^2) \subset \mathbf{R}$ such that $y_j^i \to y^i$, $t_j^i \downarrow 0$ and $x + t_j^i y_j^i \in C$ for
all $j \in \mathbf{N}$ and $i = 1, 2$. Define

$$y_j := (1 - \lambda)y_j^1 + \lambda y^2 \qquad \text{and} \qquad t_j := \min\{t_j^1, t_j^2\}.$$

Then

$$x + t_j y_j = (1 - \lambda)(x + t_j y_j^1) + \lambda(x + t_j y_j^2) \in C$$

because C is convex and

$$x + t_j y_j^i = (1 - \frac{t_j}{t_j^i})x + \frac{t_j}{t_j^i}(x + t_j^i y_j^i) \in C$$

because $\frac{t_j}{t_j^i} \in [0, 1]$ and C is convex. Thus $y_j \in T_C(x)$ and

$$\begin{aligned}
\|y_j - y\| &= \|(1 - \lambda)y_j^1 + \lambda y_j^2 - (1 - \lambda)y^1 - \lambda y^2\| \\
&\le (1 - \lambda)\|y_j^1 - y^1\| + \lambda\|y_j^2 - y^2\| \longrightarrow 0,
\end{aligned}$$

when $j \to \infty$. This means that $y \in T_C(x)$ and so $T_C(x)$ is convex. $\square$

Next we shall define the concept of normal cone. As we already have the tangent
cone, it is natural to use polarity to define the normal vectors.

Definition 2.2.5. *The normal cone of the convex set C at $x \in C$ is the set*

$$N_C(x) := T_C(x)^\circ = \{z \in \mathbf{R}^n \mid y^T z \le 0 \quad \text{for all} \quad y \in T_C(x)\} \qquad (2.9)$$

The elements of $T_C(x)$ and $N_C(x)$ are called *tangent* and *normal vectors*, respectively.

The natural corollary of this kind of duality definition is that the normal cone
has the same properties as the tangent cone.

Theorem 2.2.6. *The normal cone $N_C(x)$ of the convex set C is a closed convex
cone containing zero.*

Proof. This follows easily from the definition of $N_C(x)$. $\square$

We have the following alternative characterizations of the tangent and normal
cones.

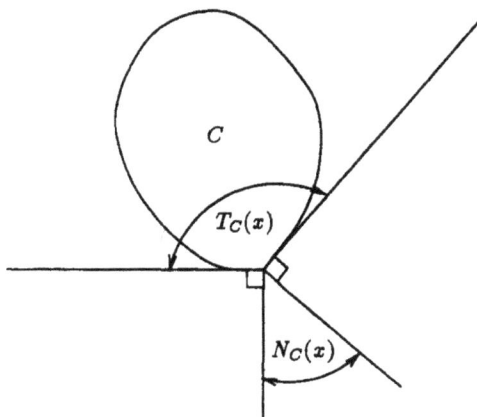

Figure 2.3. The normal cone of convex set

Theorem 2.2.7. *The tangent cone can also be written as follows*

$$T_C(x) = \mathrm{cl}\,\{y \in \mathbf{R}^n \mid \text{there exists}\ \ t > 0\ \ \text{so that}\ \ x + ty \in C\}. \qquad (2.10)$$

Proof. Let

$$K := \{y \in \mathbf{R}^n \mid \text{there exists}\ \ t > 0\ \ \text{such that}\ \ x + ty \in C\}.$$

If $y \in T_C(x)$ is arbitrary, then there exist sequences $y_j \to y$ and $t_j \downarrow 0$ such that $x + t_j y_j \in C$ for all $j \in \mathbf{N}$, so $y \in \mathrm{cl}\,K$.

To see the converse, let $y \in \mathrm{cl}\,K$. Then there exist sequences $y_j \to y$ and $t_j > 0$ such that $x + t_j y_j \in C$ for all $j \in \mathbf{N}$. It suffices now to find a sequence t'_j such that $t'_j \downarrow 0$ and $x + t'_j y_j \in C$. Choose $t'_j := \min\{\frac{1}{j}, t_j\}$, which implies that

$$|t'_j| \le \frac{1}{j} \longrightarrow 0$$

and by the convexity of C it follows that

$$x + t'_j y_j = (1 - \frac{t'_j}{t_j})x + \frac{t'_j}{t_j}(x + t_j y_j) \in C,$$

which proves the assertion. $\qquad\qquad\qquad\qquad\qquad\qquad\qquad\qquad\square$

Theorem 2.2.8. *The normal cone can also be written as follows*

$$N_C(x) = \{z \in \mathbf{R}^n \mid (x' - x)^T z \leq 0 \quad \text{for all} \quad x' \in C\}. \tag{2.11}$$

Proof. Let

$$S := \{z \in \mathbf{R}^n \mid (x' - x)^T z \leq 0 \quad \text{for all} \quad x' \in C\}.$$

Let $z \in N_C(x)$ be an arbitrary point. Then by definition

$$y^T z \leq 0 \quad \text{for all} \quad y \in T_C(x).$$

Now let $x' \in C$, set $y := x' - x$ and choose $t := 1$. Then

$$x + ty = x + tx' - tx = x' \in C,$$

so $y \in T_C(x)$. Since $z \in N_C(x)$, one has

$$(x' - x)^T z = y^T z \leq 0$$

so $z \in S$. Then it follows that

$$N_C(x) \subset S.$$

On the other hand, if $z \in S$ and $y \in T_C(x)$ then there exist sequences $(y_j) \subset \mathbf{R}^n$ and $(t_j) \subset \mathbf{R}$ such that $y_j \to y$, $t_j > 0$ and $x + t_j y_j \in C$ for all $j \in \mathbf{N}$. We set $x'_j := x + t_j y_j \in C$ so $t_j y_j^T z = (x'_j - x)^T z \leq 0$. Because t_j is positive, it implies that $y_j^T z \leq 0$ for all $j \in \mathbf{N}$. Then

$$y^T z = y_j^T z + (y - y_j)^T z$$
$$\leq y_j^T z + \|z\| \, \|y - y_j\|$$

where $\|y - y_j\| \longrightarrow 0$ as $j \to \infty$. Thus

$$y^T z \leq 0 \quad \text{for all} \quad y \in T_C(x)$$

and

$$S \subset N_C(x),$$

which completes the proof. □

2.3. Links between Geometry and Differentiation

In this section we are going to show that the analytical and geometrical concepts defined in the previous sections are actually equivalent. This equivalence is to be expressed in terms of the concepts of distance function and epigraph.

Definition 2.3.1. Let $G \subset \mathbf{R}^n$ be a nonempty set. The *distance function* $d_G :$ $\mathbf{R}^n \to \mathbf{R}$ to the set G is defined by

$$d_G(x) := \inf\{\|x - c\| \mid c \in G\} \quad \text{for all} \quad x \in \mathbf{R}^n. \tag{2.12}$$

Theorem 2.3.2. *The function d_G is Lipschitz with constant $K = 1$, in other words*

$$|d_G(x) - d_G(y)| \leq \|x - y\| \quad \text{for all} \quad x, y \in \mathbf{R}^n. \tag{2.13}$$

If the set C is convex then the function d_C is also convex.

Proof. Let any $\varepsilon > 0$ and $y \in \mathbf{R}^n$ be given. By definition, there is a point $g \in G$ such that

$$d_G(y) \geq \|y - g\| - \varepsilon.$$

Now we have

$$d_G(x) \leq \|x - g\| \leq \|x - y\| + \|y - g\|$$
$$\leq \|x - y\| + d_G(y) + \varepsilon$$

which establishes the Lipschitz condition as $\varepsilon > 0$ is arbitrary.

Suppose now that C is a convex set and let $x, y \in \mathbf{R}^n$, $\lambda \in [0, 1]$ and $\varepsilon > 0$ be given. Choose points $c_x, c_y \in C$ such that

$$\|c_x - x\| \leq d_C(x) + \varepsilon \quad \text{and} \quad \|c_y - x\| \leq d_C(y) + \varepsilon$$

and define $c := (1 - \lambda)c_x + \lambda c_y \in C$. Then

$$d_C\big((1 - \lambda)x + \lambda y\big) \leq \|c - (1 - \lambda)x - \lambda y\|$$
$$\leq (1 - \lambda)\|c_x - x\| + \lambda\|c_y - y\|$$
$$\leq (1 - \lambda)d_C(x) + \lambda d_C(y) + \varepsilon.$$

Since ε is arbitrary, d_C is convex. $\qquad\square$

Lemma 2.3.3. *If $G \subset \mathbf{R}^n$ is closed, then*

$$x \in G \iff d_G(x) = 0. \tag{2.14}$$

Proof. Let $x \in G$ be arbitrary. Then

$$0 \leq d_G(x) \leq \|x - x\| = 0$$

and thus $d_G(x) = 0$.

On the other hand if $d_G(x) = 0$, then there exists a sequence $(g_j) \subset G$ such that

$$\|x - g_j\| < 1/j \longrightarrow 0, \quad \text{when } j \to \infty.$$

Thus, the sequence (g_j) converges to x and $x \in \text{cl } G = G$. $\qquad\square$

The next two theorems show how one could equivalently define tangents and normals by using the distance function.

Theorem 2.3.4. *The tangent cone of the convex set C at $x \in C$ can also be written as*

$$T_C(x) = \{y \in \mathbf{R}^n \mid d'_C(x; y) = 0\}. \tag{2.15}$$

Proof. Let $K := \{y \in \mathbf{R}^n \mid d'_C(x; y) = 0\}$. Suppose first that $y \in T_C(x)$ is arbitrary. Then there exist sequences $(y_j) \subset \mathbf{R}^n$ and $(t_j) \subset \mathbf{R}$ such that $y_j \to y$, $t_j \downarrow 0$ and $x + t_j y_j \in C$ for all $j \in \mathbf{N}$. It is evident that $d'_C(x; y)$ is always nonnegative so it suffices to show that $d'_C(x; y) \leq 0$. We have

$$
\begin{aligned}
d'_C(x; y) &= \lim_{t \downarrow 0} \frac{d_C(x + ty) - d_C(x)}{t} \\
&= \lim_{t \downarrow 0} \frac{\inf_{c \in C} \{\|x + ty - c\|\}}{t} \\
&\leq \lim_{t \downarrow 0} \frac{\inf_{c \in C} \{\|x + ty_j - c\|\} + \|ty - ty_j\|}{t}
\end{aligned}
$$

and

$$\inf_{c \in C} \{\|x + ty_j - c\|\} = \inf_{c \in C} \{\|(1 - \tfrac{t}{t_j})x + \tfrac{t}{t_j}(x + t_j y_j) - c\|\} = 0,$$

whenever $t \leq t_j$. Therefore

$$d'_C(x; y) \leq t\|y - y_j\| \longrightarrow 0,$$

when $j \to \infty$. Thus $d'_C(x; y) = 0$ and $T_C(x) \subset K$.

Now for the converse. Let $y \in K$ and $(t_j) \subset \mathbf{R}$ be such that $t_j \downarrow 0$. By the definition of K we get

$$d'_C(x; y) = \lim_{t_j \downarrow 0} \frac{d_C(x + t_j y)}{t_j} = 0.$$

By the definition of d_C we can choose points $c_j \in C$ so that

$$\|x + t_j y - c_j\| \leq d_C(x + t_j y) + \frac{t_j}{j}.$$

Setting

$$y_j := \frac{c_j - x}{t_j},$$

we have

$$x + t_j y_j = x + t_j \frac{c_j - x}{t_j} = c_j \in C$$

and

$$\|y - y_j\| = \|y - \frac{c_j - x}{t_j}\|$$
$$= \frac{\|x + t_j y - c_j\|}{t_j}$$
$$\leq \frac{d_C(x + t_j y)}{t_j} + \frac{1}{j} \longrightarrow 0,$$

as $j \to \infty$. Thus $y \in T_C(x)$ and $K = T_C(x)$. $\qquad\qquad\square$

Theorem 2.3.5. *The normal cone of the convex set C at $x \in C$ is also given by*

$$N_C(x) = \text{cl} \left\{ \bigcup_{\lambda \geq 0} \lambda \, \partial_c d_C(x) \right\}. \tag{2.16}$$

Proof. First, let $z \in \partial_c d_C(x)$. Then by Theorem 2.1.5(ii)

$$z^T y \leq d'_C(x; y) \quad \text{for all} \quad y \in \mathbf{R}^n.$$

If one has $y \in T_C(x)$ then by Theorem 2.3.4 $d'_C(x; y) = 0$. Thus $z^T y \leq 0$ for all $y \in T_C(x)$ which implies that $z \in N_C(x)$. Because $N_C(x)$ is closed and convex cone we have

$$\text{cl} \left\{ \bigcup_{\lambda \geq 0} \lambda \, \partial_c d_C(x) \right\} \subset N_C(x).$$

For the converse, let $z \in N_C(x)$. From the definition of a normal cone and Theorem 2.3.4 we get

$$z^T y \leq 0 = d'_C(x; y) \quad \text{for all} \quad y \in T_C(x).$$

Suppose, now, that $y \notin T_C(x)$. By Theorem 2.2.4 one has $y \neq 0$ and

$$d'_C(x; y) = \lim_{t \downarrow 0} \frac{d_C(x + ty) - d_C(x)}{t} = \lim_{t \downarrow 0} \frac{1}{t} \cdot d_C(x + ty) \geq 0.$$

If $z = 0$, then we have $z^T y = 0 \leq d'_C(x; y)$. If $z \neq 0$, choose

$$\lambda_{z,y} := \frac{d'_C(x; y)}{\|z\| \, \|y\|} \geq 0.$$

Then

$$\lambda_{z,y} z^T y \leq \lambda_{z,y} \|z\| \, \|y\| = d'_C(x; y).$$

Thus, for all $z \in N_C(x)$ and $y \in \mathbf{R}^n$ we have $\lambda_{z,y} \geq 0$ such that ($\lambda_{z,y} := 1$ if $y \in T_C(x)$ or $z = 0$)

$$\lambda_{z,y} z^{\mathrm{T}} y \leq d'_C(x; y)$$

which means by Theorem 2.1.5(ii) that

$$z \in \bigcup_{\lambda \geq 0} \lambda \, \partial_c d_C(x).$$

The proof is complete.   □

We shall now define another link between geometry and differentiation: the epigraph of a given function, which is a subset of $\mathbf{R}^n \times \mathbf{R}$.

Definition 2.3.6. *The epigraph of a function $f : \mathbf{R}^n \to \mathbf{R}$ is the following subset of $\mathbf{R}^n \times \mathbf{R}$:*

$$\operatorname{epi} f := \{(x, r) \in \mathbf{R}^n \times \mathbf{R} \mid f(x) \leq r\}. \tag{2.17}$$

Theorem 2.3.7. *The epigraph of a convex function $f : \mathbf{R}^n \to \mathbf{R}$ is a closed convex subset of $\mathbf{R}^n \times \mathbf{R}$ and the epigraph of the function $v \mapsto f'(x; v)$ is a convex cone containing zero.*

Proof. This follows immediately from the definitions of convex sets and functions, and Theorem 2.1.3(ii), which says that the function $v \mapsto f'(x; v)$ is positively homogeneous and subadditive.   □

The next two theorems show how one could equivalently define tangents and normals by using the epigraph of a convex function (see Figures 2.4 and 2.5).

Theorem 2.3.8. *If the function $f : \mathbf{R}^n \to \mathbf{R}$ is convex, then*

$$\operatorname{epi} f'(x; \cdot) = T_{\operatorname{epi} f}(x, f(x)). \tag{2.18}$$

Proof. Suppose first that $(v, r) \in T_{\operatorname{epi} f}(x, f(x))$. We must show that $f'(x, v) \leq r$ which implies that $(v, r) \in \operatorname{epi} f'(x; \cdot)$. By the definition of the tangent cone there exist sequences $(v_j, r_j) \to (v, r)$ and $t_j \downarrow 0$ such that

$$(x, f(x)) + t_j(v_j, r_j) \in \operatorname{epi} f \quad \text{for all} \quad j \in \mathbf{N},$$

so that

$$f(x + t_j v_j) \leq f(x) + t_j r_j.$$

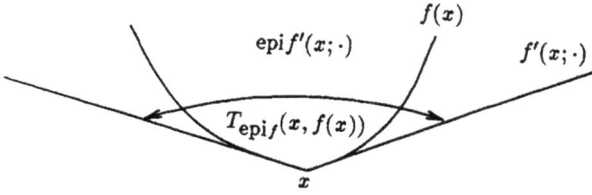

Figure 2.4.

Now we can calculate

$$f'(x;v) = \lim_{t\downarrow 0} \frac{f(x+tv) - f(x)}{t}$$

$$= \lim_{j\to\infty} \frac{f(x+t_jv_j) - f(x)}{t_j}$$

$$\le \lim_{j\to\infty} r_j = r.$$

Suppose, next, that $(v,r) \in$ epi $f'(x; \cdot)$, which means that

$$f'(x;v) = \lim_{t\downarrow 0} \frac{f(x+tv) - f(x)}{t} \le r.$$

Then there exists a sequence $t_j \downarrow 0$ such that

$$\frac{f(x+t_jv) - f(x)}{t_j} \le r + \frac{1}{j},$$

which yields

$$f(x+t_jv) \le f(x) + t_j(r + \frac{1}{j})$$

and thus $(x, f(x)) + t_j(v, r + \frac{1}{j}) \in$ epi f. This and the fact that $(v, r + \frac{1}{j}) \to (v, r)$ shows that $(v, r) \in T_{\text{epi}\, f}(x, f(x))$ and we obtain the desired conclusion. $\square$

Theorem 2.3.9. *If the function* $f : \mathbf{R}^n \to \mathbf{R}$ *is convex, then*

$$\partial_c f(x) = \{\xi \in \mathbf{R}^n \mid (\xi, -1) \in N_{\text{epi}\, f}(x, f(x))\}. \tag{2.19}$$

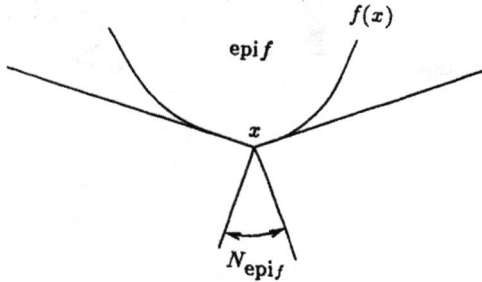

Figure 2.5.

Proof. By Theorem 2.1.5(ii) we know that ξ belongs to $\partial_c f(x)$ if and only if for any $v \in \mathbf{R}^n$ we have $f'(x; v) \geq \xi^T v$. This is equivalent to the condition that for any $v \in \mathbf{R}^n$ and $r \geq f'(x; v)$ we have $r \geq \xi^T v$, that is, for any $v \in \mathbf{R}^n$ and $r \geq f'(x; v)$ we have

$$(\xi, -1)^T (v, r) \leq 0.$$

By the definition of the epigraph and Theorem 2.3.8 we have $(v, r) \in \text{epi } f'(x; \cdot) = T_{\text{epi } f}(x; f(x))$. This and the last inequality means, by the definition of the normal cone, that $(\xi, -1)$ lies in $N_{\text{epi } f}(x; f(x))$. $\square$

Chapter 3

Nonsmooth Differential Theory

In this chapter we shall generalize the differential concepts defined in previous chapter to consider also nonconvex Lipschitz functions. The fact that for Lipschitz functions there need not exist any classical directional derivative forces us first to generalize the directional derivative. After that we generalize the other concepts analogously. The generalization can be done in many ways. We shall use the approach of **Clarke** (1983) in finite dimensional case. For other generalizations see for example **Hiriart-Urruty** (1984), **Ioffe** (1983), **Demyanov and Dixon** (1986), **Komlósi** (1989) for quasidifferentiable functions and **Ward and Borwein** (1987) for lower semicontinuous functions.

In section 3.2 we shall give several derivation rules to help the calculation of subgradients in practice. The main result of this chapter is the Theorem 3.2.16, which tells how one can compute the subdifferential by using limits of ordinary gradients. In section 3.3 we define the so called ε-subdifferential both for convex and nonconvex functions.

3.1. Generalization of Derivative

We start this section by generalizing the ordinary directional derivative. Note that this generalized derivative always exists for Lipschitz functions.

Definition 3.1.1 (Clarke). Let $f : \mathbf{R}^n \to \mathbf{R}$ be locally Lipschitz at a point $x \in \mathbf{R}^n$. The *generalized directional derivative* of f at x in the direction of $v \in \mathbf{R}^n$ is defined by

$$f^\circ(x; v) = \limsup_{\substack{y \to x \\ t \downarrow 0}} \frac{f(y + tv) - f(y)}{t}. \tag{3.1}$$

The following summarizes some basic properties of the generalized directional derivative.

Theorem 3.1.2. *Let f be locally Lipschitz at x with constant K. Then*

(i) *the function $v \mapsto f^\circ(x; v)$ is positively homogeneous and subadditive on $\mathbf{R}^n$ with*

$$|f^\circ(x; v)| \le K\|v\|,$$

29

(ii) $f°(x; v)$ is upper semicontinuous as a function of $(x; v)$ and Lipschitz with constant K as a function of v on $\mathbf{R}^n$,

(iii) $f°(x; -v) = (-f)°(x; v)$.

Proof. (i) We start by proving the inequality in assertion (i). From the Lipschitz condition we obtain

$$|f°(x; v)| = |\limsup_{\substack{y \to x \\ t \downarrow 0}} \frac{f(y + tv) - f(y)}{t}|$$

$$\leq \limsup_{\substack{y \to x \\ t \downarrow 0}} \frac{|f(y + tv) - f(y)|}{t}$$

$$\leq \limsup_{\substack{y \to x \\ t \downarrow 0}} \frac{K\|y + tv - y\|}{t},$$

whenever $y, y + tv \in B(x; \varepsilon)$. Thus

$$|f°(x; v)| \leq \limsup_{\substack{y \to x \\ t \downarrow 0}} \frac{Kt\|v\|}{t} = K\|v\|.$$

Next we show that the derivative is positively homogeneous. To see this, let $\lambda > 0$. Then

$$f°(x; \lambda v) = \limsup_{\substack{y \to x \\ t \downarrow 0}} \frac{f(y + t\lambda v) - f(y)}{t}$$

$$= \limsup_{\substack{y \to x \\ t \downarrow 0}} \lambda \cdot \left\{ \frac{f(y + t\lambda v) - f(y)}{\lambda t} \right\}$$

$$= \lambda \cdot \limsup_{\substack{y \to x \\ t \downarrow 0}} \frac{f(y + t\lambda v) - f(y)}{\lambda t} = \lambda \cdot f°(x; v).$$

We turn now to the subadditivity. Let $v, w \in \mathbf{R}^n$ be arbitrary. Then

$$f°(x; v + w) = \limsup_{\substack{y \to x \\ t \downarrow 0}} \frac{f(y + t(v + w)) - f(y)}{t}$$

$$= \limsup_{\substack{y \to x \\ t \downarrow 0}} \frac{f(y + tv + tw) - f(y + tw) + f(y + tw) - f(y)}{t}$$

$$\leq \limsup_{\substack{y \to x \\ t \downarrow 0}} \frac{f((y + tw) + tv) - f(y + tw)}{t} + \limsup_{\substack{y \to x \\ t \downarrow 0}} \frac{f(y + tw) - f(y)}{t}$$

$$= f°(x; v) + f°(x; w).$$

Thus $v \mapsto f^\circ(x; v)$ is subadditive, which establishes (i).

(ii) Let (x_i), $(v_i) \subset \mathbf{R}^n$ be sequences such that $x_i \to x$ and $v_i \to v$. By definition of upper limit, there exist sequences $(y_i) \subset \mathbf{R}^n$ and $(t_i) \subset \mathbf{R}$ such that $t_i > 0$,

$$f^\circ(x; v_i) \le [f(y_i + tv_i) - f(y_i)]/t_i + 1/i$$

and

$$\|y_i - x_i\| + t_i < 1/i \quad \text{for all} \quad i \in \mathbf{N}.$$

Now we have

$$f^\circ(x_i; v_i) - \frac{1}{i} = \limsup_{\substack{y \to x_i \\ t \downarrow 0}} \frac{f(y + tv_i) - f(y)}{t} - \frac{1}{i}$$

$$\le \frac{f(y_i + t_i v_i) - f(y_i)}{t_i}$$

$$= \frac{f(y_i + t_i v) - f(y_i)}{t_i} + \frac{f(y_i + t_i v_i) - f(y_i + t_i v)}{t_i}$$

and, by the Lipschitz condition

$$\frac{|f(y_i + t_i v_i) - f(y_i + t_i v)|}{t_i} \le \frac{K\|t_i v_i - t_i v\|}{t_i} = K\|v_i - v\| \longrightarrow 0,$$

as $i \to \infty$ provided $y_i + t_i v_i$, $y_i + t_i v \in B(x; \varepsilon)$. On taking upper limits (as $i \to \infty$), we obtain

$$\limsup_{i \to \infty} f^\circ(x_i; v_i) \le \limsup_{i \to \infty} \frac{f(y_i + t_i v) - f(y_i)}{t_i} \le f^\circ(x; v),$$

which establishes the upper semicontinuity. To show the Lipschitz condition, let $v, w \in \mathbf{R}^n$ be given. If $y + tv$ and $y + tw$ are in $B(x; \varepsilon)$, then

$$f(y + tv) - f(y + tw) \le Kt\|v - w\|,$$

and so

$$\limsup_{\substack{y \to x \\ t \downarrow 0}} \frac{f(y + tv) - f(y)}{t} \le \limsup_{\substack{y \to x \\ t \downarrow 0}} \frac{f(y + tw) - f(y)}{t} + K\|v - w\|,$$

whence

$$f^\circ(x; v) - f^\circ(x; w) \le K\|v - w\|.$$

Reversing the roles of v and w we get

$$f°(x;w) - f°(x;v) \leq K\|v - w\|.$$

Thus

$$|f°(x;v) - f°(x;w)| \leq K\|v - w\|.$$

(iii) To prove (iii) we calculate

$$f°(x;-v) = \limsup_{\substack{y \to x \\ t \downarrow 0}} \frac{f(y - tv) - f(y)}{t}$$

$$= \limsup_{\substack{u \to x \\ t \downarrow 0}} \frac{(-f)(u + tv) - (-f)(u)}{t}$$

where $u := y - tv$, and so

$$f°(x;-v) = (-f)°(x;v).$$

$\square$

We are now ready to generalize the subdifferential to nonconvex Lipschitz functions. Note that the definition is analogous to the property in Theorem 2.1.5(ii) for convex functions, with the directional derivative replaced by the generalized directional derivative.

Definition 3.1.3 (Clarke). Let $f : \mathbf{R}^n \to \mathbf{R}$ be locally Lipschitz at $x \in \mathbf{R}^n$. Then the *subdifferential* of f at x is the set

$$\partial f(x) := \{\xi \in \mathbf{R}^n \mid f°(x;v) \geq \xi^{\mathrm{T}} v \text{ for all } v \in \mathbf{R}^n\}. \tag{3.2}$$

Each element $\xi \in \partial f(x)$ is called a *subgradient* of f at x.

Some basic properties of subdifferential will be listed.

Theorem 3.1.4. *Let f be locally Lipschitz at x with constant K. Then*

(i) $\partial f(x)$ *is a nonempty, convex, compact set such that* $\partial f(x) \subset B(0; K)$,

(ii) $f°(x;v) = \max\{\xi^{\mathrm{T}} v \mid \xi \in \partial f(x)\}$ *for all $v \in \mathbf{R}^n$,*

(iii) *the mapping $\partial f(\cdot) : \mathbf{R}^n \to \mathcal{P}(\mathbf{R}^n)$ is upper semicontinuous.*

Proof. (i) In view of Theorem 3.1.2(i), the function $f^\circ(x;\cdot) : \mathbf{R}^n \to \mathbf{R}$ is positively homogeneous and subadditive. Then by the Hahn-Banach Theorem (Theorem 1.2.1), there is a vector $\xi \in \mathbf{R}^n$ such that $\xi^T v \le f^\circ(x;v)$ for all $v \in \mathbf{R}^n$, which means, by the definition of the subdifferential, that $\partial f(x)$ is nonempty. For convexity, let $\xi, \xi' \in \partial f(x)$ and $\lambda \in [0,1]$. Then

$$(\lambda \xi + (1-\lambda)\xi')^T v = \lambda \xi^T v + (1-\lambda)(\xi')^T v$$
$$\le \lambda f^\circ(x;v) + (1-\lambda)f^\circ(x;v) = f^\circ(x;v),$$

whence $\lambda \xi + (1-\lambda)\xi' \in \partial f(x)$; thus $\partial f(x)$ is convex. From Theorem 3.1.2(i) we have

$$\|\xi\|^2 = |\xi^T \xi| \le |f^\circ(x;\xi)| \le K\|\xi\|$$

and, since ξ was arbitrary we get

$$\|\xi\| \le K \quad \text{for all} \quad \xi \in \partial f(x).$$

This means that $\partial f(x)$ is bounded. For the compactness it now suffices to show that it is also closed. To see this, let $(\xi_i) \subset \partial f(x)$ be a sequence such that $\xi_i \to \xi$. Then

$$\xi^T v = \lim_{i \to \infty} \xi_i^T v \le \lim_{i \to \infty} f^\circ(x;v) = f^\circ(x;v),$$

which shows that $\xi \in \partial f(x)$ whence $\partial f(x)$ is closed.

(ii) From the definition of the subdifferential we obtain

$$f^\circ(x;v) \ge \max\{\xi^T v \mid \xi \in \partial f(x)\} \quad \text{for all} \quad v \in \mathbf{R}^n.$$

If there were $v_1 \in \mathbf{R}^n$ such that

$$f^\circ(x;v_1) > \max\{\xi^T v_1 \mid \xi \in \partial f(x)\},$$

then by the Hahn-Banach Theorem there exists $\xi_1 \in \mathbf{R}^n$ such that $f^\circ(x;v) \ge \xi_1^T v$ for all $v \in \mathbf{R}^n$ and $f^\circ(x;v_1) = \xi_1^T v_1$. It follows that $\xi_1 \in \partial f(x)$ and

$$f^\circ(x;v_1) > \max\{\xi^T v_1 \mid \xi \in \partial f(x)\} \ge \xi_1^T v_1 = f^\circ(x;v_1),$$

which is impossible. Thus

$$f^\circ(x;v) = \max\{\xi^T v \mid \xi \in \partial f(x)\} \quad \text{for all} \quad v \in \mathbf{R}^n$$

which establishes (ii).

(iii) Let $(y_i) \subset \mathbf{R}^n$ and $(\xi_i) \subset \partial f(y_i)$ for each i be given sequences such that $y_i \to x$ and $\xi_i \to \xi$. Then for all $v \in \mathbf{R}^n$ we have

$$\xi^T v = \lim_{i \to \infty} \xi_i^T v \leq \limsup_{i \to \infty} f^\circ(y_i; v).$$

By Theorem 3.1.2(ii) the function $f^\circ(x; \cdot)$ is upper semicontinuous, so

$$\xi^T v \leq f^\circ(x; v).$$

Thus the function $\partial f(\cdot)$ is also upper semicontinuous and the proof is complete. □

The next two theorems show that the subdifferential really is a generalization of the classical derivative.

Theorem 3.1.5. *Let f be locally Lipschitz at x and differentiable at x. Then*

$$\nabla f(x) \in \partial f(x). \tag{3.3}$$

Proof. By the definition of differentiability the directional derivative $f'(x; v)$ exists for all $v \in \mathbf{R}^n$ and $f'(x; v) = \nabla f(x)^T v$. Noting that $f' \leq f^\circ$, it follows that

$$f^\circ(x; v) \geq \nabla f(x)^T v \quad \text{for all} \quad v \in \mathbf{R}^n$$

and so $\nabla f(x) \in \partial f(x)$. □

Lemma 3.1.6. *If f is continuously differentiable at x, then f is locally Lipschitz at x.*

Proof. Continuous differentiability means that the linear valued derivative mapping $\nabla f : \mathbf{R}^n \to \mathcal{L}(\mathbf{R}^n, \mathbf{R})$ is continuous on a neighborhood of x. It follows that there exist constants $\varepsilon > 0$ and $M > 0$ such that

$$\|\nabla f(w)\| \leq M \quad \text{for all} \quad w \in B(x; \varepsilon).$$

Suppose now that $y, y' \in B(x; \varepsilon)$. Then, by the classical mean-value theorem, there is $z \in (y, y') \subset B(x; \varepsilon)$ such that

$$f(y) - f(y') = \nabla f(z)^T (y - y').$$

We now have

$$|f(y) - f(y')| \leq \|\nabla f(z)\| \, \|y - y'\| \leq M \, \|y - y'\|,$$

which is the Lipschitz condition at x.   □

Theorem 3.1.7. *If f is continuously differentiable at x, then*

$$\partial f(x) = \{\nabla f(x)\}. \tag{3.4}$$

Proof. In view of the Lemma 3.1.6 the function f is locally Lipschitz at x so, noting (3.3), it suffices to show the equality. The continuous differentiability means that if $x_i \to x$, then the gradient sequence $\nabla f(x_i)$ converges to $\nabla f(x)$. Now for all $v \in \mathbf{R}^n$ we calculate

$$\begin{aligned}
\lim_{x_i \to x} f'(x_i; v) &= \lim_{x_i \to x} \lim_{t \downarrow 0} \frac{f(x_i + tv) - f(x_i)}{t} \\
&= \lim_{x_i \to x} \nabla f(x_i)^{\mathrm{T}} v \\
&= \nabla f(x)^{\mathrm{T}} v = f'(x; v).
\end{aligned}$$

For all $v \in \mathbf{R}^n$, then,

$$\begin{aligned}
f'(x; v) = \lim_{x_i \to x} f'(x_i; v) = \lim_{\substack{x_i \to x \\ t \downarrow 0}} \frac{f(x_i + tv) - f(x_i)}{t} \\
= \limsup_{\substack{x_i \to x \\ t \downarrow 0}} \frac{f(x_i + tv) - f(x_i)}{t} \\
= f^{\circ}(x; v).
\end{aligned}$$

Thus $f^{\circ}(x; v) = \nabla f(x)^{\mathrm{T}} v$ for all $v \in \mathbf{R}^n$, which implies by the definition of the subdifferential that $\nabla f(x)$ is the unique subgradient at x.   □

The following shows that the subdifferential for Lipschitz functions is a generalization of the subdifferential for convex functions.

Theorem 3.1.8. *If the function $f : \mathbf{R}^n \to \mathbf{R}$ is convex, then*

(a) $f'(x; v) = f^{\circ}(x; v)$ *for all* $v \in \mathbf{R}^n$ *and*

(b) $\partial_c f(x) = \partial f(x)$.

Proof. If (a) is true, then (b) follows from the definition of subdifferential and Theorem 2.1.5(ii), so it suffices to prove (a). By the definition of the generalized directional derivative, one has $f^\circ(x; v) \geq f'(x; v)$ for all $v \in \mathbf{R}^n$. On the other hand, if $\delta > 0$ is fixed, then

$$f^\circ(x; v) = \limsup_{\substack{x' \to x \\ t \downarrow 0}} \frac{f(x' + tv) - f(x')}{t}$$

$$= \lim_{\varepsilon \downarrow 0} \sup_{\|x' - x\| < \varepsilon\delta} \sup_{0 < t < \varepsilon} \frac{f(x' + tv) - f(x')}{t}.$$

From the proof of Theorem 2.1.3 we get that the function $\varphi(t) = (1/t)(f(x' + tv) - f(x'))$ is nondecreasing and so we can write

$$f^\circ(x; v) = \lim_{\varepsilon \downarrow 0} \sup_{\|x' - x\| < \varepsilon\delta} \frac{f(x' + \varepsilon v) - f(x')}{\varepsilon}.$$

Now by the Lipschitz condition, for any $x' \in B(x; \varepsilon\delta)$, one has

$$\left| \frac{f(x' + \varepsilon v) - f(x')}{\varepsilon} - \frac{f(x + \varepsilon v) - f(x)}{\varepsilon} \right| \leq \left| \frac{f(x' + \varepsilon v) - f(x + \varepsilon v)}{\varepsilon} \right|$$

$$+ \left| \frac{f(x) - f(x')}{\varepsilon} \right|$$

$$\leq \frac{K}{\varepsilon}\|x' - x\| + \frac{K}{\varepsilon}\|x' - x\|$$

$$\leq \frac{2K}{\varepsilon}\varepsilon\delta$$

$$= 2K\delta$$

so that

$$f^\circ(x; v) \leq \lim_{\varepsilon \downarrow 0} \frac{f(x + \varepsilon v) - f(x)}{\varepsilon} + 2\delta K = f'(x; v) + 2\delta K.$$

Since $\delta > 0$ is arbitrary, we deduce

$$f^\circ(x; v) \leq f'(x; v)$$

and the proof is complete. $\qquad\qquad\qquad\qquad\qquad\qquad\qquad\qquad\qquad\qquad\qquad\square$

3.2. Subdifferential Calculus

In this section we shall derive an assortment of formulas that facilitate greatly the calculation of subdifferentials in practice. Note that all the derivation rules are generalizations of classical rules for differentiable functions. However, in this case we have to be content with inclusions instead of equalities. For this reason we shall define a regularity property, which will sharpen our rules by turning the inclusions to equalities.

Definition 3.2.1. The function $f : \mathbf{R}^n \to \mathbf{R}$ is said to be *regular* at $x \in \mathbf{R}^n$ if for all $v \in \mathbf{R}^n$ the directional derivative $f'(x; v)$ exists and

$$f'(x; v) = f^\circ(x; v). \tag{3.5}$$

We note some sufficient conditions for f to be regular.

Theorem 3.2.2. *Let f be Lipschitz at x. Then f is regular at x if*

(a) *f is continuously differentiable at x,*

(b) *f is convex,*

(c) *$f = \displaystyle\sum_{i=1}^{m} \lambda_i f_i$, where $\lambda_i > 0$ and f_i is regular at x for each $i = 1, \ldots, m$.*

Proof. (a) If f is continuously differentiable, then the directional derivative $f'(x; v)$ exists for all $v \in \mathbf{R}^n$ and by the proof of Theorem 3.1.7 $f^\circ(x; v) = f'(x; v)$ for all $v \in \mathbf{R}^n$. This gives (a).

(b) This follows from Theorems 2.1.5(ii) and 3.1.8.

(c) It suffices to prove the formula for $m = 2$; the general case follows by induction. If f is regular at x and $\lambda > 0$, then

$$(\lambda f)^\circ(x; v) = \lambda \cdot f^\circ(x; v) = \lambda \cdot f'(x; v) = (\lambda f)'(x; v) \quad \text{for all} \quad v \in \mathbf{R}^n.$$

It is evident that $(f_1 + f_2)'$ always exists and $(f_1 + f_2)' = f_1' + f_2'$. By the definition of the generalized directional derivative $(f_1 + f_2)^\circ \geq (f_1 + f_2)'$. On the other hand

$$
\begin{aligned}
(f_1 + f_2)^\circ(x; v) &= \limsup_{\substack{y \to x \\ t \downarrow 0}} \frac{(f_1 + f_2)(y + tv) - (f_1 + f_2)(y)}{t} \\
&= \limsup_{\substack{y \to x \\ t \downarrow 0}} \frac{f_1(y + tv) + f_2(y + tv) - f_1(y) - f_2(y)}{t} \\
&\leq \limsup_{\substack{y \to x \\ t \downarrow 0}} \frac{f_1(y + tv) - f_1(y)}{t} + \limsup_{\substack{y \to x \\ t \downarrow 0}} \frac{f_2(y + tv) - f_2(y)}{t} \\
&= f_1^\circ(x; v) + f_2^\circ(x; v).
\end{aligned}
$$

Thus

$$(f_1 + f_2)' = f_1' + f_2' = f_1^\circ + f_2^\circ \geq (f_1 + f_2)^\circ,$$

so

$$(f_1 + f_2)' = (f_1 + f_2)^\circ$$

and the proof is complete.   □

Corollary 3.2.3. *If f is differentiable, regular and locally Lipschitz at x, then*

$$\partial f(x) = \{\nabla f(x)\}. \tag{3.6}$$

Proof. This follows immediately from Theorem 3.2.2 and the proof of Theorem 3.1.7.   □

Theorem 3.2.4. *If f is locally Lipschitz at x, then for all $\lambda \in \mathbf{R}$*

$$\partial(\lambda f)(x) = \lambda\,\partial f(x). \tag{3.7}$$

Proof. It is evident that the function λf is also locally Lipschitz at x. If $\lambda \geq 0$ then clearly $(\lambda f)^\circ = \lambda \cdot f^\circ$, so $\partial(\lambda f)(x) = \lambda\partial f(x)$ for all $\lambda \geq 0$. It suffices now to prove the formula for $\lambda = -1$. We calculate

$$
\begin{aligned}
\xi \in \partial(-f)(x) &\iff (-f)^\circ(x;v) \geq \xi^T v \quad \text{for all} \quad v \in \mathbf{R}^n \\
&\iff f^\circ(x;-v) \geq \xi^T v \quad \text{for all} \quad v \in \mathbf{R}^n \\
&\iff f^\circ(x;-v) \geq (-\xi)^T(-v) \quad \text{for all} \quad -v \in \mathbf{R}^n \\
&\iff -\xi \in \partial f(x) \\
&\iff \xi \in -\partial f(x)
\end{aligned}
$$

and the assertion follows.   □

Theorem 3.2.5. *If f is locally Lipschitz at x and attains its extremum at x, then*

$$0 \in \partial f(x). \tag{3.8}$$

Proof. Suppose first that f attains a local minimum at x. Then there exists $\varepsilon > 0$ such that $f(x + tv) - f(x) \geq 0$ for all $0 < t < \varepsilon$ and $v \in \mathbf{R}^n$. Now we have

$$f^\circ(x;v) = \limsup_{\substack{y \to x \\ t\downarrow 0}} \frac{f(y + tv) - f(y)}{t} \geq \limsup_{t\downarrow 0} \frac{f(x + tv) - f(x)}{t} \geq 0$$

and so

$$f^\circ(x;v) \geq 0 = 0^T v \quad \text{for all} \quad v \in \mathbf{R}^n,$$

which means by the definition of subdifferential that $0 \in \partial f(x)$.

Suppose next that f attains a local maximum at x. Then $-f$ attains a local minimum at x and, as above $0 \in \partial(-f)(x)$. The statement follows then from Theorem 3.2.4. □

Theorem 3.2.6. *If the functions $f_i : \mathbf{R}^n \to \mathbf{R}$ are locally Lipschitz at x for $i = 1, \ldots, m$, then for scalars $\lambda_i \in \mathbf{R}$*

$$\partial\Big(\sum_{i=1}^{m} \lambda_i f_i\Big)(x) \subset \sum_{i=1}^{m} \lambda_i \partial f_i(x) \qquad (3.9)$$

and equality holds if, in addition, each f_i is regular at x and each $\lambda_i > 0$.

Proof. It suffices to prove the formula for $m = 2$; the general case follows by induction. In the proof of Theorem 3.2.2 we observed that

$$(f_1 + f_2)^\circ(x; v) \le f_1^\circ(x; v) + f_2^\circ(x; v),$$

whence, by the definition of subdifferential,

$$\partial(f_1 + f_2)(x) \subset \partial f_1(x) + \partial f_2(x).$$

In view of Theorem 3.2.4 we have

$$\partial(\lambda_1 f_1 + \lambda_2 f_2)(x) \subset \partial(\lambda_1 f_1)(x) + \partial(\lambda_2 f_2)(x) = \lambda_1 \partial f_1(x) + \lambda_2 \partial f_2(x).$$

Suppose next that f_i is regular at x and $\lambda_i > 0$ for $i = 1, 2$. By Theorem 3.2.2 the function $\lambda_1 f_1 + \lambda_2 f_2$ is regular; in other words

$$(\lambda_1 f_1 + \lambda_2 f_2)^\circ = (\lambda_1 f_1 + \lambda_2 f_2)' = \lambda_1 f_1' + \lambda_2 f_2' = \lambda_1 f_1^\circ + \lambda_2 f_2^\circ,$$

and it follows that

$$\partial(\lambda_1 f_1 + \lambda_2 f_2)(x) = \lambda_1 \partial f_1(x) + \lambda_2 \partial f_2(x).$$

The proof is complete. □

Next we present one of the key results of differential calculus, namely the Mean-Value Theorem. For further study of this subject we refer to **Penot** (1988), **Studniarski** (1985a, 1985b) and **Zagrodny** (1990).

Theorem 3.2.7 (Mean-Value Theorem). *Let $x, y \in \mathbf{R}^n$, $x \neq y$ and let the function f be Lipschitz on an open set $U \subset \mathbf{R}^n$ such that the line segment $[x, y] \subset U$. Then there exists a point $u \in (x, y)$ such that*

$$f(y) - f(x) \in \partial f(u)^{\mathrm{T}} (y - x). \tag{3.10}$$

Lemma 3.2.8. *The function $g : [0, 1] \to \mathbf{R}$ defined by $g(t) := f(x + t(y - x))$, is Lipschitz on $(0, 1)$ and*

$$\partial g(t) \subset \partial f(x + t(y - x))^{\mathrm{T}} (y - x). \tag{3.11}$$

Proof. We denote $x + t(y - x)$ by x_t. The function g is Lipschitz on $(0, 1)$ because

$$
\begin{aligned}
|g(t) - g(t')| = |f(x_t) - f(x_{t'})| &\leq K \, \|x_t - x_{t'}\| \\
&= K \, \|(t - t')(y - x)\| = K \, \|y - x\| \, |t - t'| \\
&= \tilde{K} \, |t - t'| \quad \text{for all} \quad t, t' \in (0, 1),
\end{aligned}
$$

where $\tilde{K} := K \, \|y - x\|$.

From Theorem 3.1.4 we get that the sets $\partial g(t)$ and $\partial f(x_t)^{\mathrm{T}} (y - x)$ are compact and convex. Since they belong to $\mathbf{R}$, they must be closed intervals in $\mathbf{R}$ and so it suffices to prove that for $v = \pm 1$, we have

$$\max \{\partial g(t) v\} \leq \max \{\partial f(x_t)^{\mathrm{T}} (y - x) v\}.$$

By Theorem 3.1.4 we have $\max \{\partial g(t) v\} = g^\circ(t; v)$ and so

$$
\begin{aligned}
\max \{\partial g(t) v\} &= \limsup_{\substack{s \to t \\ \lambda \downarrow 0}} \frac{g(s + \lambda v) - g(s)}{\lambda} \\
&= \limsup_{\substack{s \to t \\ \lambda \downarrow 0}} \frac{f(x + [s + \lambda v](y - x)) - f(x + s(y - x))}{\lambda} \\
&\leq \limsup_{\substack{v' \to x_t \\ \lambda \downarrow 0}} \frac{f(y' + \lambda v(y - x)) - f(y')}{\lambda} \\
&= f^\circ(x_t; v(y - x)).
\end{aligned}
$$

Furthermore it follows from Theorem 3.1.4 that

$$f^\circ(x_t; v(y - x)) = \max \{\partial f(x_t)^{\mathrm{T}} (v(y - x))\},$$

and thus

$$\max \{\partial g(t)v\} \leq \max \{\partial f(x_t)^{\mathrm{T}}(y - x)v\}.$$

□

Proof of the Mean-Value Theorem. Let us define the function $\Theta : [0, 1] \to \mathbf{R}$ such that $\Theta(t) := f(x_t) + t[f(x) - f(y)]$. Then it is evident that Θ is continuous and

$$\Theta(0) = f(x_0) = f(x)$$
$$\Theta(1) = f(x_1) + f(x) - f(y) = f(x).$$

Then it follows that there exists $t_0 \in (0, 1)$ such that $\Theta(t)$ attains a local extremum at t_0 and by Theorem 3.2.5 we have $0 \in \partial\Theta(t_0)$. Now by using Theorem 3.2.6 we get

$$\partial\Theta(t) = \partial[f(x_t) + t(f(x) - f(y))] \subset \partial f(x_t) + [f(x) - f(y)]\partial(t)$$

and furthermore by Lemma 3.2.8 we have

$$0 \in \partial f(x_t)^{\mathrm{T}}(y - x) + [f(x) - f(y)] \cdot \partial(t).$$

Then from the fact that $\partial(t) = 1$, it follows that

$$f(y) - f(x) \in \partial f(u)^{\mathrm{T}}(y - x),$$

where $u := x_t \in (x, y)$, which is the assertion of the theorem. □

Theorem 3.2.9 (Chain Rule). *Let $h : \mathbf{R}^n \to \mathbf{R}^m$ and $g : \mathbf{R}^m \to \mathbf{R}$ be functions such that each component function $h_i : \mathbf{R}^n \to \mathbf{R}$, $i = 1, \ldots, m$ is locally Lipschitz at $x \in \mathbf{R}^n$ and g is locally Lipschitz at $h(x) \in \mathbf{R}^m$. Then the composite function $f = g \circ h$, $f : \mathbf{R}^n \to \mathbf{R}$ is locally Lipschitz at x and*

$$\partial f(x) \subset \mathrm{conv}\left\{ \sum_{i=1}^{m} \alpha_i \xi_i \mid \xi_i \in \partial h_i(x) \quad \text{and} \quad \alpha \in \partial g(h(x)) \right\}. \tag{3.12}$$

Equality holds in (3.12) if any one of the following additional hypotheses is valid:

(i) *The function g is regular at $h(x)$, each h_i is regular at x and $\alpha_i \geq 0$ for all $i = 1, \ldots, m$. Then also f is regular at x.*

(ii) *The function g is regular at $h(x)$ and h_i is continuously differentiable at x for all $i = 1, \ldots, m$.*

(iii) *The case $m = 1$ and g is continuously differentiable at $h(x)$.*

Proof. It is evident that f is locally Lipschitz at x. Denote

$$S := \{\sum_{i=1}^{m} \alpha_i \xi_i \mid \xi_i \in \partial h_i(x) \quad \text{and} \quad \alpha \in \partial g(h(x))\}.$$

The fact that $\partial h_i(x)$ and $\partial g(h(x))$ are compact sets implies that S is also compact, and hence by Theorem 1.2.2 its convex hull is a convex compact set. Then by Theorem 1.2.3 it suffices to prove that

$$f^\circ(x; v) \leq \max_{\eta \in \text{conv} S} \eta^{\mathrm{T}} v \quad \text{for all} \quad v \in \mathbf{R}^n. \tag{3.13}$$

To see this let $\eta \in \text{conv } S$. Then we have $\eta = \sum_{j=1}^{k} \lambda^j s^j$ with $s^j \in S$, $\sum_{j=1}^{k} \lambda^j = 1$ and $\lambda^j \geq 0$ and for all $v \in \mathbf{R}^n$ one has

$$\eta^{\mathrm{T}} v = \sum_{j=1}^{k} \lambda^j (s^j)^{\mathrm{T}} v \leq \sum_{j=1}^{k} \lambda^j \max_{s \in S} s^{\mathrm{T}} v = \max_{s \in S} s^{\mathrm{T}} v$$

so

$$\max_{\eta \in \text{conv} S} \eta^{\mathrm{T}} v = \max_{s \in S} s^{\mathrm{T}} v \quad \text{for all} \quad v \in \mathbf{R}^n.$$

Define

$$q_\varepsilon(v) := \max \left\{ \sum_{i=1}^{m} \alpha_i \xi_i^{\mathrm{T}} v \mid \xi_i \in \partial h_i(x_i), \ \alpha \in \partial g(u), \ x_i \in B(x; \varepsilon), \ u \in B(h(x); \varepsilon) \right\}.$$

Then we have

$$q_0(v) = \max \left\{ \sum_{i=1}^{m} \alpha_i \xi_i^{\mathrm{T}} v \mid \xi_i \in \partial h_i(x), \ \alpha \in \partial g(h(x)) \right\}$$

$$= \max_{s \in S} s^{\mathrm{T}} v.$$

This will imply (3.13) if we show that for all $\varepsilon > 0$

$$f^\circ(x; v) - \varepsilon \leq q_\varepsilon(v) \quad \text{for all} \quad v \in \mathbf{R}^n, \tag{3.14}$$

and that $q_\varepsilon(v) \to q_0(v)$ as $\varepsilon \downarrow 0$, for all $v \in \mathbf{R}^n$.

Lemma 3.2.10. $\lim_{\varepsilon \downarrow 0} q_\varepsilon = q_0$.

Proof. To see this let $\delta > 0$ and $v \in \mathbf{R}^n$ be given. Because each h_i is locally Lipschitz at x, g is locally Lipschitz at $h(x)$ and the function $h_i^\circ(\cdot\,;\cdot)$ is upper semicontinuous by Theorem 3.1.2(ii), we can choose $\varepsilon > 0$ such that each h_i is Lipschitz on $B(x;\varepsilon)$ and g is Lipschitz on $B(h(x);\varepsilon)$ with the same constant K, and such that for all $i = 1,\ldots,m$ one has

$$h_i^\circ(x_i;\pm v) \le h_i^\circ(x;\pm v) + \delta/K \quad \text{for all} \quad x_i \in B(x;\varepsilon).$$

If $\alpha \in \partial g\,(B(h(x);\varepsilon))$ then by Theorem 3.1.4(i) we have $|\alpha_i| \le K$ for all $i = 1,\ldots,m$. By Theorem 3.1.2(i) we know that $h_i^\circ(y;\cdot)$ is positively homogeneous. Then multiplying across by $|\alpha_i|$ gives

$$h_i^\circ(x_i;\alpha_i v) \le h_i^\circ(x;\alpha_i v) + |\alpha_i|\,\delta/K \le h_i^\circ(x;\alpha_i v) + \delta.$$

On the other hand, we know by Theorem 3.1.4(iii) that the mapping $\partial g(\cdot)$ is upper semicontinuous, from which it follows that we can also choose ε small enough to guarantee that $\partial g\,(B(h(x);\varepsilon)) \subset B\,(\partial g(h(x));\delta)$. We may now calculate

$q_0 \le q_\varepsilon(v)$

$$= \max\left\{\sum_{i=1}^m \alpha_i \xi_i^{\mathrm{T}} v \mid \xi_i \in \partial h_i(x_i),\ \alpha \in \partial g(u),\ x_i \in B(x;\varepsilon),\ u \in B(h(x);\varepsilon)\right\}$$

$$\le \max\left\{\sum_{i=1}^m \max\{\alpha_i \xi_i^{\mathrm{T}} v \mid \xi_i \in \partial h_i(x_i),\ x_i \in B(x;\varepsilon)\} \mid \alpha \in B\,(\partial g(h(x));\delta)\right\}$$

$$\le \max\left\{\sum_{i=1}^m (h_i^\circ(x;\alpha_i v) + \delta) \mid \alpha \in B\,(\partial g(h(x));\delta)\right\}$$

$$\le \max\left\{\sum_{i=1}^m \max\{\alpha_i \xi_i^{\mathrm{T}} v \mid \xi_i \in \partial h_i(x)\} \mid \alpha \in B\,(\partial g(h(x));\delta)\right\} + m\delta$$

$$\le q_0 + m\delta K|v| + m\delta \longrightarrow q_0, \qquad \text{whenever} \quad \delta \to 0,$$

which completes the proof of the lemma. $\qquad\square$

Now we turn back to the proof of the chain rule. We are going to show that inequality (3.14) holds. To see this let $\varepsilon > 0$. Then by the definition of the generalized directional derivative there exist $x' \in \mathbf{R}^n$ and $t > 0$ such that

$$f^\circ(x;v) \le \frac{f(x'+tv) - f(x')}{t} + \varepsilon \tag{3.15}$$

and

$$\begin{cases} x', \ x' + tv \in B(x; \varepsilon) \\ h(x'), h(x' + tv) \in B(h(x); \varepsilon). \end{cases}$$

By the Mean-Value Theorem 3.2.7 there exists $\alpha \in \partial g(u)$ such that $u \in [h(x' + tv), h(x')] \subset B(h(x); \varepsilon)$ and

$$f(x' + tv) - f(x') = g(h(x' + tv)) - g(h(x')) = \alpha^T(h(x' + tv) - h(x'))$$
$$= \sum_{i=1}^{m} \alpha_i[h_i(x' + tv) - h_i(x')].$$

We apply the Mean-Value Theorem again to the functions h_i, $i = 1, \ldots, m$. Then there exist subgradients $\xi_i \in \partial h_i(x_i)$ such that $x_i \in [x' + tv, x'] \subset B(x; \varepsilon)$ and

$$f(x' + tv) - f(x') = \sum_{i=1}^{m} \alpha_i[h_i(x' + tv) - h_i(x')] = \sum_{i=1}^{m} \alpha_i \xi_i^T(x' + tv - x')$$
$$= \sum_{i=1}^{m} \alpha_i \xi_i^T(tv).$$

Now it follows from (3.15) that

$$f^\circ(x; v) \leq \frac{f(x' + tv) - f(x')}{t} + \varepsilon = \frac{\sum_{i=1}^{m} \alpha_i \xi_i^T(tv)}{t} + \varepsilon$$
$$= \frac{t \sum_{i=1}^{m} \alpha_i \xi_i^T v}{t} + \varepsilon = \sum_{i=1}^{m} \alpha_i \xi_i^T v + \varepsilon \qquad (3.16)$$
$$\leq q_\varepsilon(v) + \varepsilon \quad \text{for all} \quad v \in \mathbf{R}^n,$$

which establishes (3.14) and the basic assertion.

We turn now to the additional hypotheses (i)-(iii).

(i) Suppose first that g is regular at $h(x)$, each h_i is regular at x and $\alpha_i > 0$ for all $i = 1, \ldots, m$. To prove the equality in (3.12) it suffices to show that

$$f^\circ(x; v) = q_0(v) \quad \text{for all} \quad v \in \mathbf{R}^n.$$

From above we found that $f^\circ(x; v) \leq q_0(v)$ for all $v \in \mathbf{R}^n$. On the other hand, the fact that $\alpha_i \geq 0$ for all $i = 1, \ldots, m$, h_i is regular at x and g is regular at $h(x)$

imply

$$q_0(v) = \max \left\{ \sum_{i=1}^m \alpha_i \xi_i^T v \mid \xi_i \in \partial h_i(x), \ \alpha \in \partial g(h(x)) \right\}$$

$$\leq \max \left\{ \sum_{i=1}^m \alpha_i \max_{\xi_i \in \partial h_i(x)} \xi_i^T v \mid \alpha \in \partial g(h(x)) \right\}$$

$$= \max \left\{ \sum_{i=1}^m \alpha_i h_i^\circ(x; v) \mid \alpha \in \partial g(h(x)) \right\}$$

$$= \max \left\{ \sum_{i=1}^m \alpha_i h_i'(x; v) \mid \alpha \in \partial g(h(x)) \right\}$$

$$= g^\circ(h(x); h'(x; v)) = g'(h(x); w),$$

where $w_i := h_i'(x; v)$. Then by definition the directional derivative

$$g'(h(x); w) = \lim_{t \downarrow 0} \frac{g(h(x) + tw) - g(h(x))}{t}$$

$$= \lim_{t \downarrow 0} \left\{ \frac{g(h(x + tv)) - g(h(x))}{t} + T \right\},$$

where $T := g(h(x) + tw) - g(h(x + tv))/t$. We obtain an upper estimate of T and show that it goes to zero, when $t \to 0$. Due to the Lipschitz property of the function g one has

$$T \leq \frac{|g(h(x) + tw) - g(h(x + tv))|}{t} \leq \frac{K \|h(x) + tw - h(x + tv)\|}{t}$$

$$= K \left\| w - \frac{h(x + tv) - h(x)}{t} \right\| \longrightarrow K \|h'(x, v) - h'(x, v)\| = 0,$$

as $t \to 0$. Thus,

$$q_0(v) \leq \lim_{t \downarrow 0} \frac{g(h(x + tv)) - g(h(x))}{t} = f'(x; v) \leq f^\circ(x; v)$$

and by (3.16) we have $q_0(v) = f'(x; v) = f^\circ(x; v)$ for all $v \in \mathbf{R}^n$. In other words, f is regular at x and equality holds in (3.12), which establishes (i).

(ii) Suppose next that the function g is regular at $h(x)$ and each h_i is continuously differentiable at x for all $i = 1, \dots, m$. Then by Theorem 3.1.7 we have

$$q_0(v) = \max \left\{ \sum_{i=1}^{m} \alpha_i \xi_i^T v \mid \xi_i \in \partial h_i(x) = \{\nabla h_i(x)\}, \ \alpha \in \partial g(h(x)) \right\}$$

$$= \max \left\{ \sum_{i=1}^{m} \alpha_i \nabla h_i(x)^T v \mid \alpha \in \partial g(h(x)) \right\}$$

$$= \max \left\{ \sum_{i=1}^{m} \alpha_i h_i'(x, v) \mid \alpha \in \partial g(h(x)) \right\}.$$

Now we can continue in the same way as in (i) to get

$$q_0(v) \le f^\circ(x; v),$$

which proves assertion (ii).

(iii) Finally, suppose $m = 1$ and that the function g is continuously differentiable at $h(x)$. Then

$$\alpha = g'(h(x)) = \lim_{y \to x} \frac{g(h(x)) - g(h(y))}{h(x) - h(y)}$$

and $\lim_{x' \to x} g'(h(x')) = \alpha$. We may assume that $\alpha \ge 0$. Then we calculate

$$q_0(v) = \max \{\alpha \xi^T v \mid \xi \in \partial h(x)\} = \alpha \cdot h^\circ(x; v)$$

$$= \limsup_{\substack{x' \to x \\ t \downarrow 0}} \frac{\alpha[h(x' + tv) - h(x')]}{t} = \limsup_{\substack{x' \to x \\ t \downarrow 0}} \frac{g'(h(x'))[h(x' + tv) - h(x')]}{t}$$

$$= \limsup_{\substack{x' \to x \\ t \downarrow 0}} \frac{g(h(x' + tv)) - g(h(x'))}{t} = f^\circ(x; v) \quad \text{for all} \quad v \in \mathbf{R}^n$$

and the theorem is proved. $\square$

Theorem 3.2.11 (Products). *Let f_1 and f_2 be locally Lipschitz at $x \in \mathbf{R}^n$. Then the function $f_1 f_2$ is locally Lipschitz at x and*

$$\partial(f_1 f_2)(x) \subset f_2(x) \partial f_1(x) + f_1(x) \partial f_2(x). \tag{3.17}$$

If in addition $f_1(x), f_2(x) \ge 0$ and f_1, f_2 are both regular at x, then equality holds and the function $f_1 f_2$ is regular at x.

Proof. Define the function $g : \mathbf{R}^2 \to \mathbf{R}$ by $g(u_1, u_2) := u_1 \cdot u_2$. Then g is continuously differentiable and by Theorem 3.1.7 it is locally Lipschitz at $(f_1(x), f_2(x))$ with

$$\partial g(f_1(x), f_2(x)) = \{\nabla g(f_1(x), f_2(x))\} = \{(f_2(x), f_1(x))\}.$$

Next define the function $h : \mathbf{R}^n \to \mathbf{R}^2$ by $h(x) := (f_1(x), f_2(x))$. Now we have $f_1 \cdot f_2 = g \circ h$. By the chain rule the function $f_1 \cdot f_2$ is locally Lipschitz at x and

$$\partial(f_1 f_2)(x) \subset \text{conv}\Big\{ \sum_{i=1}^{2} \alpha_i \xi_i \mid \xi_i \in \partial h_i(x), \ \alpha \in \partial g(h(x)) \Big\}$$
$$= \text{conv}\{ f_2(x)\partial f_1(x) + f_1(x)\partial f_2(x) \}.$$

Then by Theorems 3.1.4(i) and 2.1.2 the set $f_2(x)\partial f_1(x) + f_1(x)\partial f_2(x)$ is convex and so

$$\partial(f_1 f_2)(x) \subset f_2(x)\partial f_1(x) + f_1(x)\partial f_2(x).$$

Suppose next that $f_1(x), f_2(x) \geq 0$ and that f_1, f_2 are regular at x. The function g is regular by Theorem 3.2.2(i). Then by Theorem 3.2.9(i) the function $f_1 f_2$ is regular at x and equality holds in (3.17). $\square$

Theorem 3.2.12 (Quotients). Let f_1 and f_2 be locally Lipschitz at $x \in \mathbf{R}^n$ and $f_2(x) \neq 0$. Then the function f_1/f_2 is locally Lipschitz at x and

$$\partial\Big(\frac{f_1}{f_2}\Big)(x) \subset \frac{f_2(x)\partial f_1(x) - f_1(x)\partial f_2(x)}{f_2^2(x)}. \tag{3.18}$$

If in addition $f_1(x) \geq 0$, $f_2(x) > 0$ and f_1, f_2 are both regular at x, then equality holds and the function f_1/f_2 is regular at x.

Proof. The proof is nearly the same as for products. $\square$

The following theorem deals with a class of functions which are frequently encountered in nonsmooth optimization, namely max-functions. The problem of minimizing such a function is usually called the min-max problem.

Theorem 3.2.13. Let $f_i : \mathbf{R}^n \to \mathbf{R}$ be locally Lipschitz at x for each $i = 1, \ldots, m$. Then the function $f : \mathbf{R}^n \to \mathbf{R}$ defined by

$$f(x) := \max \{ f_i(x) \mid i = 1, \ldots, m \}$$

is locally Lipschitz at x and

$$\partial f(x) \subset \text{conv} \{ \partial f_i(x) \mid i \in I(x) \}, \tag{3.19}$$

where $I(x) = \{i \in \{1, \ldots, m\} \mid f_i(x) = f(x)\}$. In addition if f_i is regular at x for each $i = 1, \ldots, m$, then f is regular at x and equality holds in (3.19).

Proof. The function f is evidently locally Lipschitz at x. Define $g : \mathbf{R}^m \to \mathbf{R}$ and $h : \mathbf{R}^n \to \mathbf{R}^m$ by

$$g(u_1, \ldots, u_m) := \max_{i=1,\ldots,m} \{u_i\}$$

$$h(x) := \big(f_1(x), \ldots, f_m(x)\big).$$

Now we have $f = g \circ h$. For all $u, v \in \mathbf{R}^m$ and $\lambda \in [0, 1]$ it holds

$$\begin{aligned}
g(\lambda u + (1 - \lambda)v) &= \max_{i=1,\ldots,m} \{\lambda u_i + (1 - \lambda)v_i\} \\
&\leq \lambda \max_{i=1,\ldots,m} \{u_i\} + (1 - \lambda) \max_{i=1,\ldots,m} \{v_i\} \\
&= \lambda g(u) + (1 - \lambda)g(v),
\end{aligned}$$

which means that g is convex and by Theorem 2.1.2 locally Lipschitz at $h(x)$. Let $J(u) = \{i \in \{1, \ldots, m\} \mid u_i = g(u)\}$. Then the directional derivative is

$$\begin{aligned}
g'(u; v) &= \lim_{t \downarrow 0} \frac{g(u + tv) - g(u)}{t} = \lim_{t \downarrow 0} \max_{i=1,\ldots,m} \frac{\{u_i + tv_i\} - g(u)}{t} \\
&= \lim_{t \downarrow 0} \max_{i \in J(u)} \frac{\{u_i + tv_i\} - g(u)}{t} = \lim_{t \downarrow 0} \max_{i \in J(u)} \frac{\{u_i + tv_i - u_i\}}{t}.
\end{aligned}$$

Thus

$$g'(u; v) = \max_{i \in J(u)} v_i$$

and by Theorem 3.1.8 we have $g^\circ = g'$, which gives

$$\partial g(u) = \{\alpha \in \mathbf{R}^m \mid \max_{i \in J(u)} v_i \geq \alpha^{\mathrm{T}} v \quad \text{for all} \quad v \in \mathbf{R}^m\}.$$

Now it is easy to see that

$$\alpha \in \partial g(u) \iff \begin{cases} \alpha_i \geq 0, \quad i = 1, \ldots, m, \\ \displaystyle\sum_{i=1}^{m} \alpha_i = 1, \\ \alpha_i = 0, \quad \text{when } i \notin J(u) \end{cases}$$

and so we can calculate the subdifferential at $h(x) \in \mathbf{R}^m$ by

$$\partial g(h(x)) = \Big\{\alpha \in \mathbf{R}^m \mid \alpha_i \geq 0, \sum_{i=1}^{m} \alpha_i = 1 \quad \text{and} \quad \alpha_i = 0 \quad \text{if } i \notin I(x)\Big\}.$$

By applying Theorem 3.2.9 to f we get

$$\partial f(x) \subset \text{conv}\left\{\sum_{i=1}^{m} \alpha_i \xi_i \mid \xi_i \in \partial h_i(x) \quad \text{and} \quad \alpha \in \partial g(h(x))\right\}$$

$$= \text{conv}\left\{\sum_{i \in I(x)} \alpha_i \partial f_i(x) \mid \alpha_i \geq 0 \quad \text{and} \quad \sum_{i \in I(x)} \alpha_i = 1\right\}$$

$$= \text{conv}\left\{\partial f_i(x) \mid i \in I(x)\right\}.$$

Suppose next that f_i is, in addition, regular at x for all $i \in I(x)$. Because g is convex, it is, by Theorem 3.2.2(b), also regular at $h(x)$. Then the fact that $\alpha_i \geq 0$ for all $\alpha \in \partial g(h(x))$ and Theorem 3.2.9(i) imply that f is regular at x and equality holds in (3.19). □

Corollary 3.2.14. *Suppose that the functions $f_i : \mathbf{R}^n \to \mathbf{R}$ are continuously differentiable at x and $g_i : \mathbf{R}^n \to \mathbf{R}$ are convex for each $i = 1, \ldots, m$. Define the functions $f : \mathbf{R}^n \to \mathbf{R}$ and $g : \mathbf{R}^n \to \mathbf{R}$ by*

$$f(x) = \max\left\{f_i(x) \mid i = 1, \ldots, m\right\} \quad \text{and}$$

$$g(x) = \max\left\{g_i(x) \mid i = 1, \ldots, m\right\}.$$

Then we have

$$\partial f(x) = \text{conv}\left\{\nabla f_i(x) \mid i \in I(x)\right\} \quad \text{and}$$

$$\partial_c g(x) = \text{conv}\left\{\partial_c g_i(x) \mid i \in J(x)\right\},$$

(3.20)

where $I(x) = \{i \in \{1, \ldots, m\} \mid f_i(x) = f(x)\}$ and $J(x) = \{i \in \{1, \ldots, m\} \mid g_i(x) = g(x)\}$.

Proof. This follows from Theorems 3.1.7, 3.2.2 and 3.2.13. □

Theorem 3.2.15 (Rademacher). *Let $U \subset \mathbf{R}^n$ be an open set and $f : U \to \mathbf{R}$ be Lipschitz of rank K on U. Then f is differentiable almost everywhere on U.*

Proof. In view of the Lipschitz condition for all $v \in \mathbf{R}^n$ one has

$$\limsup_{t \downarrow 0}\left|\frac{f(x + tv) - f(x)}{t}\right| \leq \limsup_{t \downarrow 0}\frac{tK\|v\|}{t} = K\|v\| < +\infty.$$

Then it follows from Theorem 1.2.4 that the function f is differentiable almost everywhere on U. □

Now we are ready to present the main result of this chapter. By Rademacher's Theorem we know that a Lipschitz function is differentiable almost everywhere and thus the gradient exists almost everywhere. Let us remind that Ω_f denotes the set where f is not differentiable.

Theorem 3.2.16. Let $f : \mathbf{R}^n \to \mathbf{R}$ be locally Lipschitz at $x \in \mathbf{R}^n$. Then

$$\partial f(x) = \mathrm{conv}\left\{\xi \in \mathbf{R}^n \mid \exists\, (x_i) \subset \mathbf{R}^n \setminus \Omega_f \text{ s.t. } x_i \to x \text{ and } \nabla f(x_i) \to \xi\right\}. \quad (3.21)$$

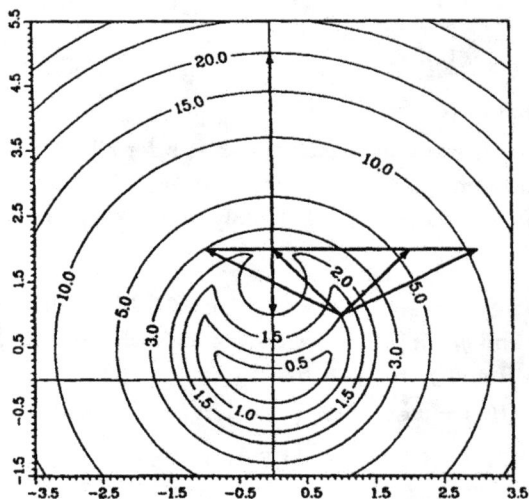

Figure 3.1. The subdifferential of nonconvex function

Proof. Let

$$A := \left\{\xi \in \mathbf{R}^n \mid \exists\, (x_i) \subset \mathbf{R}^n \setminus \Omega_f \text{ s.t. } x_i \to x \text{ and } \nabla f(x_i) \to \xi\right\}.$$

We shall first show that A is nonempty. It follows from Rademacher's Theorem that the measure of the set Ω_f is zero. Then there exists a sequence $(x_i) \subset \mathbf{R}^n$ such that $x_i \notin \Omega_f$ and $x_i \to x$. Since f is locally Lipschitz at x, Theorem 3.1.4(i) implies that there exists $\varepsilon > 0$ such that for all $x_i \in B(x; \varepsilon)$

$$\|\xi_i\| \leq K \quad \text{for all} \quad \xi_i \in \partial f(x_i). \quad (3.22)$$

This means that the point-to-set mapping ∂f is locally bounded on $B(x; \varepsilon)$. By Theorem 3.1.5 we have

$$\nabla f(x_i) \in \partial f(x_i)$$

and so by (3.22) the sequence $\{\nabla f(x_i)\}$ is bounded. Then by the Bolzano-Weierstrass Theorem (Theorem 1.2.5) the sequence $\{\nabla f(x_i)\}$ admits a convergent subsequence $\{\nabla f(x_{i_k})\}$ and so there exists $\xi \in \mathbf{R}^n$ such that $\nabla f(x_{i_k}) \to \xi$. From

Theorem 3.1.4(iii) we know that ∂f is upper semicontinuous which means that $\xi \in \partial f(x)$.

Now we have proved that A is nonempty, bounded set and $A \subset \partial f(x)$. By Theorem 3.1.4(i) the set $\partial f(x)$ is convex and so we have

$$\operatorname{conv} A \subset \operatorname{conv} \partial f(x) = \partial f(x).$$

For the reverse inclusion we show that A is also closed, hence compact. To see this, let $(\xi_j) \in A$ be a sequence such that $\xi_j \to \xi$. Then

$$\xi_j = \lim_{i \to \infty} \nabla f(x_i^j), \qquad \text{where } x_i^j \to x \text{ as } i \to \infty \text{ and } x_i^j \notin \Omega_f.$$

Extracting subsequences if necessary, there are points $x_i \in \mathbf{R}^n$ such that

$$x_i := \lim_{j \to \infty} x_i^j \quad \text{for each} \quad i \in \mathbf{N}.$$

Then it holds that $x_i \to x$, $x_i \notin \Omega_f$ and

$$\xi = \lim_{j \to \infty} \lim_{i \to \infty} \nabla f(x_i^j) = \lim_{i \to \infty} \lim_{j \to \infty} \nabla f(x_i^j) = \lim_{i \to \infty} \nabla f(x_i).$$

Thus $\xi \in A$ and so the set A is closed and compact. This implies by Theorem 1.2.2 that its convex hull $\operatorname{conv} A$ is also compact. Then we only need to show that

$$f^\circ(x; v) = \max_{\xi \in \partial f(x)} \xi^{\mathrm{T}} v \le \max_{\eta \in \operatorname{conv} A} \eta^{\mathrm{T}} v \quad \text{for all} \quad v \in \mathbf{R}^n.$$

This follows from the next lemma.

Lemma 3.2.17. *Let $\varepsilon > 0$ and $v \in \mathbf{R}^n$ be such that $v \ne 0$. Then*

$$f^\circ(x; v) - \varepsilon \le \limsup \{\nabla f(y)^{\mathrm{T}} v \mid y \to x, \ y \notin \Omega_f\}. \tag{3.23}$$

Proof. Let $\alpha := \limsup \{\nabla f(y)^{\mathrm{T}} v \mid y \to x, \ y \notin \Omega_f\}$. Then, by definition, there exists $\delta > 0$ such that the conditions

$$y \in B(x; \delta) \qquad \text{and} \qquad y \notin \Omega_f$$

imply $\nabla f(y)^{\mathrm{T}} v \le \alpha + \varepsilon$. We also choose δ small enough so that Ω_f has measure zero in $B(x; \delta)$. Now consider the line segments $L_y = \{y + tv \mid 0 < t < \delta/(2|v|)\}$. Since Ω_f has measure zero in $B(x; \delta)$, it follows from Fubini's Theorem (Theorem

1.2.6) that for almost every y in $B(x; \delta/2)$, the line segment L_y meets Ω_f in a set of zero one-dimensional measure. Let y be any point in $B(x; \delta/2)$ having this property and let t lie in $\big(0, \delta/(2|v|)\big)$. Then

$$f(y + tv) - f(y) = \int_0^t \nabla f(y + sv)^T v \, ds,$$

since ∇f exists almost everywhere on L_y. Since one has $|y + sv - x| < \delta$ for $0 < s < t$, it follows that $\nabla f(y + sv)^T v \leq \alpha + \varepsilon$, whence

$$f(y + tv) - f(y) \leq t(\alpha + \varepsilon).$$

Since this is true for almost all y within $\delta/2$ of x and for all t in $\big(0, \delta/(2|v|)\big)$, and since f is continuous, it is in fact true for all such y and t. We deduce that

$$f^\circ(x; v) \leq \alpha + \varepsilon,$$

which completes the proof. $\square$

3.3. $\mathcal{E}$-subdifferentials

In nonsmooth optimization, so-called bundle methods are based on the theory of ε-subdifferential, which is a modification of the ordinary subdifferential. Therefore we shall now give the definition and present some of its basic properties.

3.3.1. Convex Case

First we shall generalize the ordinary directional derivative.

Definition 3.3.1.1. Let $f : \mathbf{R}^n \to \mathbf{R}$ be convex. The ε-*directional derivative* of f at x in the direction of $v \in \mathbf{R}^n$ is defined by

$$f_\varepsilon'(x; v) = \inf_{t > 0} \frac{f(x + tv) - f(x) + \varepsilon}{t}. \tag{3.24}$$

Now we can reach the same results as in Theorem 2.1.3 also for the ε-directional derivative. Note the analogy with the property (2.2).

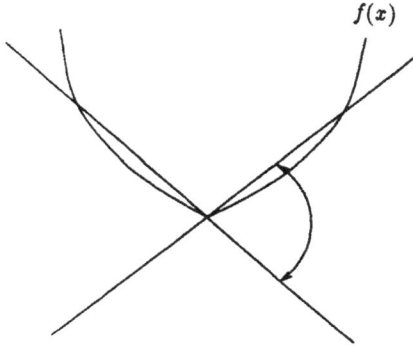

Figure 3.2.

Theorem 3.3.1.2. *Let* $f : \mathbf{R}^n \to \mathbf{R}$ *be convex. Then at every point* $x \in \mathbf{R}^n$

(i) *the function* $v \mapsto f'_\varepsilon(x; v)$ *is positively homogeneous and subadditive on* $\mathbf{R}^n$
with

$$|f'_\varepsilon(x; v)| \le K \|v\|,$$

(ii) $f'_\varepsilon(x; v)$ *is upper semicontinuous as a function of* $(x; v)$ *and Lipschitz with constant* K *as a function of* v *on* $\mathbf{R}^n$,

(iii) $-f'_\varepsilon(x; -v) \le f'_\varepsilon(x; v)$.

Proof. These results follow immediately from Theorem 2.1.3 and the fact that for all $\varepsilon > 0$ we have $\inf_{t>0} \varepsilon/t = 0$. □

As before we now generalize the subgradient and the subdifferential of a convex function. We illustrate the meaning of the definition in Figure 3.2.

Definition 3.3.1.3. *Let* $\varepsilon \ge 0$, *then the* ε-*subdifferential of the convex function* $f : \mathbf{R}^n \to \mathbf{R}$ *at* $x \in \mathbf{R}^n$ *is the set*

$$\partial_\varepsilon f(x) = \{\xi \in \mathbf{R}^n \mid f(x') \ge f(x) + \xi^{\mathrm{T}}(x' - x) - \varepsilon \quad \text{for all} \quad x' \in \mathbf{R}^n\}. \tag{3.25}$$

Each element $\xi \in \partial_\varepsilon f(x)$ *is called an* ε-*subgradient of the convex function* f *at* x.

The following summarizes some basic properties of the ε-subdifferential.

Theorem 3.3.1.4. *Let* $f : \mathbf{R}^n \to \mathbf{R}$ *be convex. Then*

(i) $\partial_0 f(x) = \partial_c f(x)$.

(ii) *If* $\varepsilon_1 \leq \varepsilon_2$, *then* $\partial_{\varepsilon_1} f(x) \subset \partial_{\varepsilon_2} f(x)$.

(iii) $f'_\varepsilon(x; v) = \max \{\xi^T v \mid \xi \in \partial_\varepsilon f(x)\}$ *for all* $v \in \mathbf{R}^n$.

(iv) $\partial_\varepsilon f(x) = \{\xi \in \mathbf{R}^n \mid f'_\varepsilon(x; v) \geq \xi^T v$ *for all* $v \in \mathbf{R}^n\}$.

(v) $\partial_\varepsilon f(x)$ *is nonempty, convex and compact set such that* $\|\xi\| \leq K$ *for all* $\xi \in \partial_\varepsilon f(x)$.

(vi) *The mapping* $\partial_\varepsilon f(\cdot) : \mathbf{R}^n \to \mathcal{P}(\mathbf{R}^n)$ *is upper semicontinuous.*

Proof. The definition of the ε-subdifferential implies directly the assertions (i) and (ii) and the proofs of assertions (iii) and (iv) are the same as for $\varepsilon = 0$ in Theorem 2.1.5. By Theorem 2.1.5(iii) $\partial_c f(x)$ is nonempty which implies by assertion (i) that $\partial_\varepsilon f(x)$ is also nonempty. The proofs of the convexity, compactness and upper semicontinuity are also same as in Theorem 2.1.5. $\square$

The following shows that the ε-subdifferential contains in a condensed form the subgradient information from a whole neighborhood.

Theorem 3.3.1.5. *Let* $f : \mathbf{R}^n \to \mathbf{R}$ *be convex with Lipschitz constant K at x. If* $\varepsilon \geq 0$, *then*

$$\partial_c f(y) \subset \partial_\varepsilon f(x) \qquad \text{for all} \quad y \in B\left(x; \tfrac{\varepsilon}{2K}\right). \tag{3.26}$$

Proof. Let $\xi \in \partial_c f(y)$ and $y \in B\left(x; \tfrac{\varepsilon}{2K}\right)$. Then for all $z \in \mathbf{R}^n$ it holds

$$f(z) \geq f(y) + \xi^T(z - y)$$
$$= f(x) + \xi^T(z - x) - (f(x) - f(y) + \xi^T(z - x) - \xi^T(z - y))$$

and, using the Lipschitz condition and Theorem 2.1.5(iii), we calculate

$$|f(x) - f(y) + \xi^T(z - x) - \xi^T(z - y)| \leq |f(x) - f(y)| + |\xi^T(z - x) - \xi^T(z - y)|$$
$$\leq K \|x - y\| + \|\xi\| \|x - y\|$$
$$\leq 2K \|x - y\| \leq 2K \cdot \frac{\varepsilon}{2K} = \varepsilon,$$

which gives $\xi \in \partial_\varepsilon f(x)$. $\square$

3.3.2. Nonconvex Case

Now we turn back to the nonconvex case. It would be possible to generalize the ε-subdifferential for convex functions analogously also for Lipschitz functions by using the generalized directional derivative. However, the theory of nonsmooth optimization has shown that it will be more useful to use the Goldstein ε-subdifferential for nonconvex functions. For this reason we shall now define and examine that instead of the natural generalization.

Definition 3.3.2.1. Let $\varepsilon \geq 0$, then the *Goldstein ε-subdifferential of the Lipschitz function* $f : \mathbf{R}^n \to \mathbf{R}$ at $x \in \mathbf{R}^n$ is the set

$$\partial_\varepsilon^G f(x) = \text{conv}\,\{\partial f(y) \mid y \in B(x;\varepsilon)\}. \tag{3.27}$$

Each element $\xi \in \partial_\varepsilon^G f(x)$ is called an *ε-subgradient at x of the function f.*

The following theorem summarizes some basic properties of the Goldstein ε-subdifferential.

Theorem 3.3.2.2. Let $f : \mathbf{R}^n \to \mathbf{R}$ be locally Lipschitz of rank K at x. Then

(i) $\partial_0^G f(x) = \partial f(x)$.

(ii) If $\varepsilon_1 \leq \varepsilon_2$, then $\partial_{\varepsilon_1}^G f(x) \subset \partial_{\varepsilon_2}^G f(x)$.

(iii) $\partial_\varepsilon^G f(x)$ is a nonempty, convex, compact set such that $\|\xi\| \leq K$ for all $\xi \in \partial_\varepsilon^G f(x)$.

(iv) The mapping $\partial_\varepsilon^G f(\cdot) : \mathbf{R}^n \to \mathcal{P}(\mathbf{R}^n)$ is upper semicontinuous.

Proof. Assertions (i) and (ii) follow directly from definition of the ε-subdifferential. We turn now to (iii). Assertion (i) implies that for all $\varepsilon \geq 0$

$$\partial f(x) = \partial_0^G f(x) \subset \partial_\varepsilon^G f(x).$$

From Theorem 3.1.4(i) we know that $\partial f(x)$ is nonempty and so $\partial_\varepsilon^G f(x)$ is also nonempty. Because $\partial_\varepsilon^G f(x)$ is the convex hull of a set, it is evidently convex and the compactness follows from the same property of $\partial f(x)$. Next we shall show the inequality. Let $\xi \in \partial_\varepsilon^G f(x)$ be arbitrary. Then $\xi = \sum_{i=1}^m \lambda_i \xi_i$, where $\xi_i \in \partial f(y_i)$, $y_i \in B(x;\varepsilon)$, $\lambda_i \geq 0$ and $\sum_{i=1}^m \lambda_i = 1$. Now by Theorem 3.1.4(i) we have

$$\|\xi\| = \|\sum_{i=1}^m \lambda_i \xi_i\| \leq \sum_{i=1}^m \lambda_i \|\xi_i\| \leq \sum_{i=1}^m \lambda_i \cdot K = K.$$

Assertion (iv) follows directly from the same property of $\partial f(x)$ and so the proof is complete.   □

As a corollary to Theorem 3.2.16 we get the next result.

Corollary 3.3.2.3. *Let* $f : \mathbf{R}^n \to \mathbf{R}$ *be locally Lipschitz at* $x \in \mathbf{R}^n$. *Then*

$$\partial_\varepsilon^G f(x) = \text{conv} \left\{ \xi \in \mathbf{R}^n \mid \exists \, (y_i) \subset \mathbf{R}^n \setminus \Omega_f \text{ s.t. } y_i \to y, \right.$$
$$\left. \nabla f(y_i) \to \xi \quad \text{and} \quad y \in B(x; \varepsilon) \right\}. \tag{3.28}$$

Proof. This follows directly from Theorem 3.2.16.   □

The following result is analogous to Theorem 3.3.1.5.

Corollary 3.3.2.4. *Let* $f : \mathbf{R}^n \to \mathbf{R}$ *be locally Lipschitz at* x. *If* $\varepsilon \geq 0$, *then*

$$\partial f(y) \subset \partial_\varepsilon^G f(x) \quad \text{for all} \quad y \in B(x; \varepsilon). \tag{3.29}$$

Proof. This follows directly from definition of the Goldstein ε-subdifferential. □

We conclude this chapter by considering the relationship between the ε-subdifferential for convex functions and the Goldstein ε-subdifferential.

Theorem 3.3.2.5. *Let* $f : \mathbf{R}^n \to \mathbf{R}$ *be convex with Lipschitz constant* K *at* x. *Then for all* $\varepsilon \geq 0$ *we have*

$$\partial_\varepsilon^G f(x) \subset \partial_{2K\varepsilon} f(x). \tag{3.30}$$

Proof. Suppose, that $\xi \in \partial_\varepsilon^G f(x)$. Then $\xi = \sum_{i=1}^m \lambda_i \xi_i$, where $\xi_i \in \partial f(y_i)$, $y_i \in B(x; \varepsilon)$, $\lambda_i \geq 0$ and $\sum_{i=1}^m \lambda_i = 1$. Since $\xi_i \in \partial f(y_i)$, for all $i = 1, \ldots, m$ one has

$$f(x') \geq f(y_i) + \xi_i^T(x' - y_i) \quad \text{for all} \quad x' \in \mathbf{R}^n.$$

We multiply both sides by λ_i and sum over i to get

$$f(x') = \sum_{i=1}^m \lambda_i f(x') \geq \sum_{i=1}^m \lambda_i f(y_i) + \sum_{i=1}^m \lambda_i \xi_i^T(x' - y_i) \quad \text{for all} \quad x' \in \mathbf{R}^n$$

$$= f(x) + \xi^T(x' - x) - \left(f(x) - \sum_{i=1}^m \lambda_i f(y_i) + \xi^T(x' - x) - \sum_{i=1}^m \lambda_i \xi_i^T(x' - y_i) \right)$$

and by using the Lipschitz condition and Theorem 3.1.4(i) we calculate

$$|f(x) - \sum_{i=1}^{m} \lambda_i f(y_i) + \xi^{\mathrm{T}}(x' - x) - \sum_{i=1}^{m} \lambda_i \xi_i^{\mathrm{T}}(x' - y_i)|$$

$$\leq |f(x) - \sum_{i=1}^{m} \lambda_i f(y_i)| + |\sum_{i=1}^{m} \lambda_i \xi^{\mathrm{T}}(x' - x) - \sum_{i=1}^{m} \lambda_i \xi_i^{\mathrm{T}}(x' - y_i)|$$

$$\leq \sum_{i=1}^{m} \lambda_i |f(x) - f(y_i)| + \sum_{i=1}^{m} \lambda_i |\xi_i^{\mathrm{T}}(x - y_i)|$$

$$\leq \sum_{i=1}^{m} \lambda_i (K \|x - y_i\| + \|\xi\| \|x - y_i\|) = 2K\varepsilon,$$

which means that $\xi \in \partial_{2K\varepsilon} f(x)$.   □

3.4. Generalized Jacobians

In part III we shall need derivatives of a nonsmooth vector-valued function F : $\mathbf{R}^n \to \mathbf{R}^m$, written in terms of component functions as $F(x) = (F_1(x), \ldots, F_m(x))^{\mathrm{T}}$. Each component function F_i for $i = 1, \ldots, m$ (and hence F) is supposed to be locally Lipschitz. Then, due to the Rademacher's Theorem 3.2.15 we conclude that F is differentiable almost everywhere. We denote again by Ω_F the set in $\mathbf{R}^n$ where F fails to be differentiable and by $\nabla F(x)$ for $x \notin \Omega_F$ the usual $m \times n$ Jacobian matrix. Based on Theorem 3.2.16 we generalize now the derivative of F.

Definition 3.4.1. Let $F : \mathbf{R}^n \to \mathbf{R}^m$ be locally Lipschitz at a point $x \in \mathbf{R}^n$. Then the *generalized Jacobian* of F at x is the set

$$\partial F(x) := \mathrm{conv} \left\{ Z \in \mathbf{R}^{m \times n} \mid \exists (x_i) \subset \mathbf{R}^n \setminus \Omega_F \right. \tag{3.31}$$
$$\left. \text{s.t. } x_i \to x \quad \text{and} \quad \nabla F(x_i) \to Z \right\}.$$

The space of $m \times n$ matrices is endowed with the norm

$$\|Z\|_{m \times n} := \left(\sum_{i=1}^{m} \|Z_i\|^2 \right)^{\frac{1}{2}}, \tag{3.32}$$

where Z_i is the ith row of Z. Some basic properties of $\partial F(x)$ will now be listed.

Corollary 3.4.2. *Let F_i for $i = 1, \ldots, m$ be locally Lipschitz at x with constant K_i. Then*

(i) $F(x) = (F_1(x), \ldots, F_m(x))^{\mathrm{T}}$ *is locally Lipschitz at x with constant $K = \|(K_1, \ldots, K_m)^{\mathrm{T}}\|$,*

(ii) *$\partial F(x)$ is a nonempty, convex, compact subset of $\mathbf{R}^{m \times n}$,*

(iii) *the mapping $\partial F(\cdot) : \mathbf{R}^n \to \mathcal{P}(\mathbf{R}^n)$ is upper semicontinuous.*

Proof. Follows directly from Theorems 3.1.4 and 3.2.16. $\qquad\qquad\square$

The following chain rule will be exploited in part III.

Corollary 3.4.3 (Jacobian Chain Rule). *Let $f = g \circ F$, where $F : \mathbf{R}^n \to \mathbf{R}^m$ is locally Lipschitz at $x \in \mathbf{R}^n$ and $g : \mathbf{R}^m \to \mathbf{R}$ is locally Lipschitz at $F(x) \in \mathbf{R}^m$. Then f is locally Lipschitz at x and*

$$\partial f(x) \subset \operatorname{conv} \left\{ \partial F(x)^{\mathrm{T}} \partial g(F(x)) \right\}. \tag{3.33}$$

If in addition g is continuously differentiable, then $\partial g(F(x)) = \{\nabla g(F(x))\}$ and

$$\partial f(x) = \partial F(x)^{\mathrm{T}} \partial g(F(x)). \tag{3.34}$$

Proof. Follows from Chain Rule 3.2.9 and Theorem 3.2.16. $\qquad\qquad\square$

Chapter 4

Nonsmooth Geometry

Now we shall turn back to geometry. We are going to show in this chapter that the geometrical concepts can also be analogously generalized in nonconvex analysis. We shall again follow the approach of **Clarke** (1983), but also here we can define the tangent and normal cones in many ways (see for example **Penot** (1984) and **Ward** (1988)).

In section 4.1 we define tangents and normals for nonconvex sets and in section 4.2 we link these concepts to the analytic ones as in the convex case.

4.1. Generalization of Tangents and Normals

First we define the tangent by using the distance function. Note that this definition is based on Theorem 2.3.4.

Definition 4.1.1 (Clarke). Let G be a nonempty subset of $\mathbf{R}^n$. *The tangent cone of the set G at $x \in G$ is given by the formula*

$$T_G(x) := \{y \in \mathbf{R}^n \mid d_G^\circ(x; y) = 0\}. \tag{4.1}$$

The tangent cone now has the same elementary properties as in the convex case.

Theorem 4.1.2. *The tangent cone $T_G(x)$ of the set G is a closed convex cone containing zero.*

Proof. This follows directly from Theorems 3.1.2(i) and (ii). □

Definition 4.1.3. Let G be a nonempty subset of $\mathbf{R}^n$. *The normal cone of the set G at $x \in G$ is the set*

$$N_G(x) := T_G(x)^\circ = \{z \in \mathbf{R}^n \mid y^{\mathrm{T}}z \leq 0 \quad \text{for all} \quad y \in T_G(x)\}. \tag{4.2}$$

The normal cone now has the same properties as before.

59

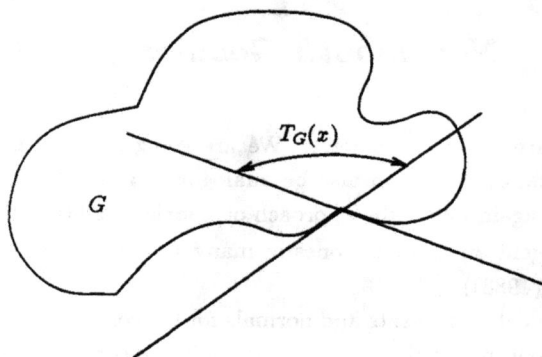

Figure 4.1. The tangent cone of nonconvex set

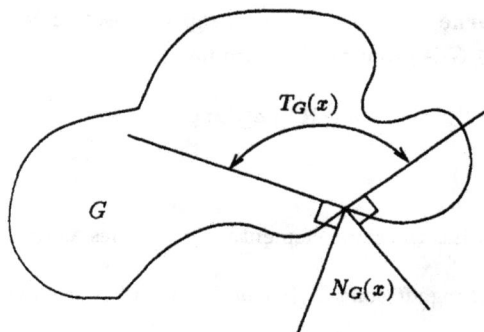

Figure 4.2. The normal cone of nonconvex set

Theorem 4.1.4. *The normal cone $N_G(x)$ of the set G is a closed convex cone containing zero.*

Proof. This follows easily from the definition of $N_G(x)$.   □

The following shows that we have actually defined the generalizations of the corresponding convex concepts.

Theorem 4.1.5. *If the set G is convex, then the tangent cone and normal cone of G as defined above coincide with the tangent cone and normal cone, respectively, defined earlier for convex sets.*

Proof. Suppose G is convex and $x \in G$. Then by Theorem 2.3.2 the function d_G is convex and by Theorem 3.1.8(a) we have

$$d_G'(x; y) = d_G^\circ(x; y) \qquad \text{for all} \quad y \in \mathbf{R}^n.$$

Then the assertion of the theorem follows from Theorem 2.3.4, from which the tangent cone of the convex set G at x can be written as

$$T_G(x) = \{y \in \mathbf{R}^n \mid d_G'(x; y) = 0\}.$$

$\square$

We have the following alternative characterizations of the tangent and normal cones. Note that they are similar to the ones in Chapter 2.

Theorem 4.1.6. *The tangent cone at x of the set G can also be written as*

$$T_G(x) = \{y \in \mathbf{R}^n \mid \text{for all } t_j \downarrow 0 \text{ and } x_j \to x \text{ with } x_j \in G,$$
$$\text{there exists } y_j \to y \text{ with } x_j + t_j y_j \in G\}. \tag{4.3}$$

Proof. Let

$$S := \{y \in \mathbf{R}^n \mid \text{for all } t_j \downarrow 0 \text{ and } x_j \to x \text{ with } x_j \in G,$$
$$\text{there exists } y_j \to y \text{ with } x_j + t_j y_j \in G\}.$$

Suppose first that $y \in T_G(x)$, and that sequences $x_j \to x$ with $x_j \in G$ and $t_j \downarrow 0$ are given. Since $d_G^\circ(x; y) = 0$ and $x_j \in G$ by assumption, we have

$$0 \le \lim_{j \to \infty} \frac{d_G(x_j + t_j y)}{t_j} = \lim_{j \to \infty} \frac{d_G(x_j + t_j y) - d_G(x_j)}{t_j}$$
$$\le \limsup_{\substack{x' \to x \\ t \downarrow 0}} \frac{d_G(x' + ty) - d_G(x')}{t} = d_G^\circ(x; y) = 0.$$

It follows that the limit exists and is zero. Then for all $j \in \mathbf{N}$ there exists $g_j \in G$ such that

$$\|x_j + t_j y - g_j\| \le d_G(x_j + t_j y) + \frac{t_j}{j};$$

we now define

$$y_j = \frac{g_j - x_j}{t_j}.$$

Then

$$\|y - y_j\| = \left\| y - \frac{g_j - x_j}{t_j} \right\|$$

$$= \frac{\|x_j + t_j - g_j\|}{t_j} \le \frac{d_G(x_j + t_j y)}{t_j} + \frac{1}{j} \longrightarrow 0$$

as $j \to \infty$ and

$$x_j + t_j y_j = x_j + t_j \left(\frac{g_j - x_j}{t_j} \right) = g_j \in G,$$

so $y \in S$.

Now for the converse. Suppose that $y \in S$ and choose sequences $z_j \to x$ and $t_j \downarrow 0$ such that

$$\lim_{j \to \infty} \frac{d_G(z_j + t_j y) - d_G(z_j)}{t_j} = d_G^\circ(x; y). \tag{4.4}$$

To prove that $d_G^\circ(x; y) = 0$, it suffices to show that the quantity in (4.4) is non-positive. To see this, choose $g_j \in G$ such that

$$\|g_j - z_j\| \le d_G(z_j) + \frac{t_j}{j}.$$

Then,

$$\|x - g_j\| \le \|x - z_j\| + \|z_j - g_j\| \le \|x - z_j\| + d_G(z_j) + \frac{t_j}{j} \longrightarrow 0$$

as $j \to \infty$. Then by assumption there exists a sequence y_j converging to y such that $g_j + t_j y_j \in G$. By Theorem 2.3.2 the function d_G is Lipschitz and we get

$$d_G(z_j + t_j y) \le d_G(g_j + t_j y_j) + \|z_j - g_j\| + t_j \|y - y_j\|$$

$$\le d_G(z_j) + t_j \left(\|y - y_j\| + \frac{1}{j} \right).$$

This implies that the quantity in (4.4) is nonpositive, which completes the proof.

$\square$

Theorem 4.1.7. *The normal cone at x of the set G can also be written as*

$$N_G(x) = \mathrm{cl} \left\{ \bigcup_{\lambda \ge 0} \lambda \, \partial d_G(x) \right\}. \tag{4.5}$$

Proof. Let $z \in \partial d_G(x)$ be given. Then, by the definition of the subdifferential,

$$z^T y \leq d_G^\circ(x; y) \quad \text{for all} \quad y \in \mathbf{R}^n.$$

If one has $y \in T_G(x)$ then, by the definition of the tangent cone, $d_G^\circ(x; y) = 0$. Thus, $z^T y \leq 0$ for all $y \in T_G(x)$ which implies that $z \in T_G(x)^\circ = N_G(x)$. Because $N_G(x)$ is a closed convex cone we have

$$\text{cl} \left\{ \bigcup_{\lambda \geq 0} \lambda \, \partial d_G(x) \right\} \subset N_G(x).$$

For the converse, let $z \in N_G(x)$. From the definitions of normal and tangent cones we get

$$z^T y \leq 0 = d_C^\circ(x; y) \quad \text{for all} \quad y \in T_G(x).$$

Suppose now that $y \notin T_G(x)$; then, by Theorem 4.1.2, $y \neq 0$. It follows that

$$d_G^\circ(x; y) = \limsup_{\substack{x' \to x \\ t \downarrow 0}} \frac{d_G(x' + ty) - d_G(x')}{t} \geq \limsup_{t \downarrow 0} \frac{1}{t} \cdot d_G(x + ty) \geq 0.$$

If $z = 0$, then we have $z^T y = 0 \leq d_G^\circ(x; y)$ while if $z \neq 0$ we choose

$$\lambda_{z,y} := \frac{d_G^\circ(x; y)}{\|z\| \, \|y\|} \geq 0$$

and have

$$\lambda_{z,y} z^T y \leq \lambda_{z,y} \|z\| \, \|y\| = d_G^\circ(x; y).$$

Thus for all $z \in N_G(x)$ and $y \in \mathbf{R}^n$ we have $\lambda_{z,y} \geq 0$ such that ($\lambda_{z,y} := 1$ if $y \in T_G(x)$ or $z = 0$)

$$\lambda_{z,y} z^T y \leq d_G^\circ(x; y),$$

which means that

$$z \in \bigcup_{\lambda \geq 0} \lambda \, \partial d_G(x)$$

and the proof is complete. $\square$

In the following chapter we shall need the next two properties of tangents and normals.

Theorem 4.1.8. *If* $x \in \operatorname{int} G$, *then*

$$T_G(x) = \mathbf{R}^n \quad \text{and} \quad N_G(x) = \{0\}. \tag{4.6}$$

Proof. Let $y \in \mathbf{R}^n$ be arbitrary. If $x \in \operatorname{int} G$, then there exists $\delta > 0$ such that $B(x; \delta) \subset G$. Choose sequences $x_j \to x$ and $t_j \downarrow 0$ such that

$$d_G^\circ(x; y) = \limsup_{j \to \infty} \frac{d_G(x_j + t_j y) - d_G(x_j)}{t_j}.$$

Then there exists $j_0 \in \mathbf{N}$ such that x_j, $x_j + t_j y \in B(x; \delta) \subset G$ for all $j \geq j_0$. Thus we have $d_G(x_j) = d(x_j + t_j y) = 0$ for all $j \geq j_0$, so $d^\circ(x; y) = 0$ and thus $y \in T_G(x)$. Now if $z \in N_G(x)$, then by definition we have

$$y^T z \leq 0 \quad \text{for all} \quad y \in T_G(x) = \mathbf{R}^n.$$

Applying this property for any $y \in \mathbf{R}^n$ and $-y \in \mathbf{R}^n$, we get $z = 0$ and the proof is complete. □

Theorem 4.1.9. *If* G_1 *and* $G_2 \subset \mathbf{R}^n$ *are such that* $x \in G_1 \cap G_2$ *and* $x \in \operatorname{int} G_2$, *then*

$$T_{G_1}(x) = T_{G_1 \cap G_2}(x) \quad \text{and} \quad N_{G_1}(x) = N_{G_1 \cap G_2}(x). \tag{4.7}$$

Proof. Let $y \in \mathbf{R}^n$ be arbitrary. Suppose that $x \in G_1 \cap G_2$ and $x \in \operatorname{int} G_2$. Then there exists $\delta > 0$ such that $B(x; \delta) \subset G_2$. It is evident that d_{G_1} and $d_{G_1 \cap G_2}$ coincides in $B(x; \delta)$ and thus $d_{G_1}^\circ(x; y) = d_{G_1 \cap G_2}^\circ(x; y)$ for all $y \in \mathbf{R}^n$. Then the assertions of the theorem follow from the definitions of tangent and normal cone. □

4.2. Epigraphs and Level Sets of Nonsmooth Functions

In this section we shall present corresponding results to those of Chapter 2 concerning the concept of epigraph.

Theorem 4.2.1. *If the function* $f : \mathbf{R}^n \to \mathbf{R}$ *is locally Lipschitz at* x, *then the epigraph of the function* $v \mapsto f^\circ(x; v)$ *is a convex cone containing zero.*

Proof. By Theorem 3.1.2(i) the function $v \mapsto f^\circ(x; v)$ is subadditive and positively homogeneous, which implies the assertion. □

Theorem 4.2.2. *If the function* $f : \mathbf{R}^n \to \mathbf{R}$ *is locally Lipschitz at* x, *then*

$$\text{epi } f^\circ(x; \cdot) = T_{\text{epi } f}(x, f(x)). \tag{4.8}$$

Proof. Suppose first that $(v, r) \in T_{\text{epi } f}(x, f(x))$. We must show that $f^\circ(x, v) \le r$ implies that $(v, r) \in \text{epi } f^\circ(x; \cdot)$. To see this, choose sequences $x_j \to x$ and $t_j \downarrow 0$ such that

$$\lim_{j \to \infty} \frac{f(x_j + t_j v) - f(x_j)}{t_j} = \limsup_{\substack{y \to x \\ t \downarrow 0}} \frac{f(y + tv) - f(y)}{t} = f^\circ(x; v).$$

By the Lipschitz condition we have

$$\|(x_j, f(x_j)) - (x, f(x))\|^2 = \|x_j - x\|^2 + |f(x_j) - f(x)|^2$$
$$\le (1 + K^2)\|x_j - x\|^2 \longrightarrow 0,$$

as $j \to \infty$, which means that the sequence $(x_j, f(x_j)) \in \text{epi } f$ is converging to $(x, f(x))$. By Theorem 4.1.6 there exists a sequence (v_j, r_j) converging to (v, r) such that

$$(x_j, f(x_j)) + t_j(v_j, r_j) \in \text{epi } f \quad \text{for all} \quad j \in \mathbf{N},$$

so

$$f(x_j + t_j v_j) \le f(x_j) + t_j r_j.$$

Now we can calculate

$$f^\circ(x; v) = \lim_{j \to \infty} \frac{f(x_j + t_j v_j) - f(x_j)}{t_j}$$
$$\le \lim_{j \to \infty} r_j = r.$$

Suppose next that $(v, r) \in \text{epi } f^\circ(x; \cdot)$, which means that $f^\circ(x; v) \le r$. Define $\delta \ge 0$ such that

$$f^\circ(x; v) + \delta = r$$

and let $t_j \downarrow 0$ and $(x_j, s_j) \in \text{epi } f$ be arbitrary sequences such that $(x_j, s_j) \to (x, f(x))$. Define sequences $v_j := v$ and

$$r_j := \max\{f^\circ(x; v) + \delta, \frac{f(x_j + t_j v) - f(x_j)}{t_j}\}.$$

Then the fact that

$$\limsup_{j \to \infty} \frac{f(x_j + t_j v) - f(x_j)}{t_j} \le f^\circ(x; v)$$

shows that $r_j \to f^\circ(x; v) + \delta = r$ and, since $(x_j, s_j) \in$ epi f, we have

$$s_j + t_j r_j \geq s_j + [f(x_j + t_j v) - f(x_j)] \geq f(x_j) - f(x_j) + f(x_j + t_j v),$$

which means that $(x_j, s_j) + t_j(v_j, r_j) \in$ epi f. Now Theorem 4.1.6 implies that $(v, r) \in T_{\text{epi } f}(x, f(x))$ and we obtain the desired conclusion. □

Theorem 4.2.3. *If the function* $f : \mathbf{R}^n \to \mathbf{R}$ *is locally Lipschitz at* x, *then*

$$\partial f(x) = \{\xi \in \mathbf{R}^n \mid (\xi, -1) \in N_{\text{epi } f}(x, f(x))\}. \tag{4.9}$$

Proof. By the definition of the subdifferential we know that ξ belongs to $\partial f(x)$ if and only if, for any $v \in \mathbf{R}^n$ we have $f^\circ(x; v) \geq \xi^T v$. This is equivalent to the condition that for any $v \in \mathbf{R}^n$ and $r \geq f^\circ(x; v)$ we have $r \geq \xi^T v$, that is, for any $v \in \mathbf{R}^n$ and $r \geq f^\circ(x; v)$ we have

$$(\xi, -1)^T (v, r) \leq 0.$$

By the definition of the epigraph and Theorem 4.2.2 we have $(v, r) \in$ epi $f^\circ(x; \cdot) = T_{\text{epi } f}(x; f(x))$. This and the last inequality mean, by the definition of the normal cone, that $(\xi, -1)$ lies in $N_{\text{epi } f}(x; f(x))$. □

Definition 4.2.4. *The* level set *of the function* $f : \mathbf{R}^n \to \mathbf{R}$ *at* x *is the set*

$$\text{lev } f(x) := \{y \in \mathbf{R}^n \mid (y, f(x)) \in \text{epi } f\}.$$

To the end of this section we give the relationships between the level sets and epigraphs.

Theorem 4.2.5. *Suppose the function* $f : \mathbf{R}^n \to \mathbf{R}$ *is locally Lipschitz at* x *and* $0 \notin \partial f(x)$. *Then*

$$\{v \in \mathbf{R}^n \mid (v, 0) \in \text{epi } f^\circ(x; \cdot)\} \subset T_{\text{lev } f(x)}(x). \tag{4.10}$$

If in addition f *is regular at* x, *then equality holds in* (4.10).

Proof. Let $S := \{v \in \mathbf{R}^n \mid (v, 0) \in \text{epi } f^\circ(x; \cdot)\}$. Then $v \in S$ if and only if $f^\circ(x; v) \leq 0$. Suppose first that $v \in S$ such that

$$f^\circ(x; v) = \limsup_{\substack{y \to x \\ t \downarrow 0}} \frac{f(y + tv) - f(y)}{t} < 0.$$

Then there exist $\varepsilon > 0$ and $\delta > 0$ such that

$$\frac{f(y + tv) - f(y)}{t} < -\delta \qquad \text{for all} \quad y \in B(x; \varepsilon) \quad \text{and} \quad t \in (0, \varepsilon). \qquad (4.11)$$

Let $x_j \to x$ and $t_j \downarrow 0$ be arbitrary sequences such that $x_j \in \text{lev } f(x)$. Then there exists $j_0 \in \mathbb{N}$ such that $x_j \in B(x; \varepsilon)$ and $t_j \in (0, \varepsilon)$ for all $j \geq j_0$ and, by the definition of the set $\text{lev } f(x)$, one has $f(x_j) \leq f(x)$. By (4.11) we have for all $j \geq j_0$

$$f(x_j + t_j v) \leq f(x_j) - \delta t_j \leq f(x) - \delta t_j \leq f(x),$$

so $x_j + t_j v \in \text{lev } f(x)$ for all $j \geq j_0$. Then, setting $v_j := v$, we deduce from Theorem 4.1.6 that $v \in T_{\text{lev } f(x)}(x)$.

Suppose next that $v \in S$ such that $f^\circ(x; v) = 0$. If there were $f^\circ(x; v) \geq 0$ for all $v \in \mathbb{R}^n$, then by the definition of the subdifferential one would have $0 \in \partial f(x)$, contradicting the assertion. Thus there always exists $\hat{v} \in S$ such that $f^\circ(x; \hat{v}) < 0$. Now define the sequence $v_j := v + \frac{1}{j}\hat{v}$. Then $v_j \to v$ and by Theorem 3.1.2(i)

$$f^\circ(x; v_j) = f^\circ(x; v + \frac{1}{j}\hat{v}) \leq f^\circ(x; v) + \frac{1}{j}f^\circ(x; \hat{v}) < 0,$$

so, as at the beginning of the proof, $v_j \in T_{\text{lev } f(x)}(x)$. By Theorem 4.1.2 $T_{\text{lev } f(x)}(x)$ is closed, which implies that $v \in T_{\text{lev } f(x)}(x)$.

Finally, suppose that f is regular at x and $v \in T_{\text{lev } f(x)}(x)$. Then the directional derivative at x exists for v and $f'(x; v) = f^\circ(x; v)$. Let $t_j \downarrow 0$. Then by Theorem 4.1.6 there exists a sequence $v_j \to v$ such that $x + t_j v_j \in \text{lev } f(x)$, so $f(x + t_j v_j) \leq f(x)$ for all $j \in \mathbb{N}$. By the Lipschitz condition we get

$$\frac{f(x + t_j v) - f(x)}{t_j} = \frac{f(x + t_j v) - f(x + t_j v_j) + f(x + t_j v_j) - f(x)}{t_j}$$

$$\leq \frac{f(x + t_j v_j) - f(x)}{t_j} + \frac{K\|x + t_j v - x - t_j v_j\|}{t_j}$$

$$\leq 0 + K\|v - v_j\|,$$

whence taking the limit as $j \to \infty$, one has $f^\circ(x; v) = f'(x; v) \leq 0$. Then $(v, 0) \in \text{epi } f^\circ(x; \cdot)$ and the proof is complete. $\qquad \square$

Theorem 4.2.6. *If the function* $f : \mathbb{R}^n \to \mathbb{R}$ *is locally Lipschitz at* x *such that* $0 \notin \partial f(x)$. *Then*

$$N_{\text{lev } f(x)}(x) \subset \bigcup_{\lambda \geq 0} \lambda \partial f(x). \qquad (4.12)$$

If, in addition, f is regular at x, then equality holds in (4.12).

Proof. Let $z \in N_{\text{lev } f(x)}(x)$, whence by the definition of the normal cone, $z^T v \leq 0$ for all $v \in T_{\text{lev } f(x)}(x)$. Then by Theorem 4.2.5

$$z^T v \leq 0 \qquad \text{for all} \quad v \in \mathbf{R}^n \quad \text{such that} \quad f^\circ(x; v) \leq 0. \qquad (4.13)$$

Suppose first that $f^\circ(x; v) > 0$. If $z^T v \leq 0$, we can choose $\lambda := 1$ and we get

$$\lambda z^T v = z^T v \leq f^\circ(x; v).$$

If $z^T v > 0$ we choose $0 \leq \lambda \leq f^\circ(x; v)/z^T v$, from which it follows that

$$(\lambda z)^T v = \lambda z^T v \leq f^\circ(x; v).$$

Suppose next that $f^\circ(x, v) = 0$. Then by (4.13) $z^T v \leq 0$ and we can choose $\lambda := 1$ to get $(\lambda z)^T v \leq f^\circ(x; v)$.

Suppose finally that $f^\circ(x, v) < 0$. If $z^T v < 0$ we choose again $0 \leq \lambda \leq f^\circ(x; v)/z^T v$, which implies that

$$\lambda z^T v = \lambda z^T v \leq f^\circ(x; v).$$

The last case is that $f^\circ(x, v) < 0$ and $z^T v = 0$. Then we consider the sequence $z_j := z - \frac{1}{j} v$, for which $z_j \to z$ and

$$z_j^T v = z - \frac{1}{j} v^T v = z^T v - \frac{1}{j} v^T v = -\frac{1}{j} \|v\|^2 < 0.$$

By the previous case there exists $\lambda \geq 0$ such that

$$\lambda z_j^T v \leq f^\circ(x; v).$$

Now by the definition of the subdifferential we have shown that

$$z \in \text{cl} \Big\{ \bigcup_{\lambda \geq 0} \lambda \partial f(x) \Big\}.$$

To complete the first part of the proof we must show that the set $S := \cup_{\lambda \geq 0} \lambda \partial f(x)$ is closed. To see this let $(z_j) \subset S$ be a sequence such that $z_j \to z$. We are going to prove that $z \in S$. The fact that $z_j \in S$ shows that $z_j = \lambda_j \xi_j$ for all $j \in \mathbf{N}$ where $\lambda_j \geq 0$ and $\xi_j \in \partial f(x)$. By Theorem 3.1.4(i) the sequence ξ_j is bounded, so there exists a subsequence $\xi_{j_i} \subset \partial f(x)$ such that $\xi_{j_i} \to \xi$. Because $\partial f(x)$ is closed, it follows that $\xi \in \partial f(x)$. Further, since $0 \notin \partial f(x)$ one has $\xi \neq 0$, so the sequence λ_{j_i} is converging to some $\lambda \geq 0$. Then $\lambda_{j_i}\xi_{j_i} \to \lambda\xi = z$, which means that $z \in S$ and so S is closed.

If in addition f is regular at x and $y \in S$, then $y = \lambda\xi$, where $\lambda \geq 0$ and $\xi \in \partial f(x)$. This implies that $\xi^T v \leq f^{\circ}(x; v)$ for all $v \in \mathbf{R}^n$. Let $v \in T_{\mathrm{lev}\, f(x)}(x)$. Then by Theorem 4.2.5 $f^{\circ}(x; v) \leq 0$, so

$$y^T v = \lambda\xi^T v \leq \lambda f^{\circ}(x; v) \leq 0$$

whence $y \in N_{\mathrm{lev}\, f(x)}(x)$ and the proof is complete. $\qquad\qquad \square$

Chapter 5

Nonsmooth Optimization Theory

In this chapter we are going to present some results connecting the theories of nonsmooth analysis and optimization. First we generalize the classical first order optimality conditions for both unconstrained and constrained optimization. For further study of first order optimality conditions we refer to **Studniarski** (1989a) and **Schirotzek** (1986).

We shall give the necessary conditions for a Lipschitz function to attain its local minimum. For convex functions these conditions are also sufficient and the minimum is global. After that we linearize the unconstrained and constrained optimization problems by using the subgradient information. These linearizations have proved to be very suitable for function approximation in nonsmooth optimization.

5.1. Optimality Conditions

We start by giving the basic necessary conditions for an unconstrained optimization problem at its minimizer x. Note that for convex functions the condition is also sufficient for x to be a minimizer.

Theorem 5.1.1. *If $f : \mathbf{R}^n \to \mathbf{R}$ is locally Lipschitz at x and attains its local minimum at x, then*

(i) $0 \in \partial f(x)$ *and*

(ii) $f^\circ(x; v) \geq 0$ *for all $v \in \mathbf{R}^n$.*

Proof. This follows directly from the proof of Theorem 3.2.5. $\square$

Theorem 5.1.2. *If $f : \mathbf{R}^n \to \mathbf{R}$ is convex, then the following conditions are equivalent:*

(i) *f attains its global minimum at x,*

(ii) $0 \in \partial_c f(x)$,

(iii) $f'(x; v) \geq 0$ *for all $v \in \mathbf{R}^n$.*

Proof. The assertion (ii) follows from (i) by Theorem 5.1.1(i). Suppose next, that (ii) holds. By Theorem 2.1.5(i) we have for all $v \in \mathbf{R}^n$

$$f'(x;v) = \max \{\xi^T v \mid \xi \in \partial_c f(x)\} \geq 0^T v = 0.$$

To complete the proof it suffices now to show that (iii) implies (i). To see this, consider $x' \in \mathbf{R}^n$. We have

$$f'(x;x'-x) = \max \{\xi^T(x'-x) \mid \xi \in \partial_c f(x)\}$$

and so there exists $\xi_{x'} \in \partial_c f(x)$ such that

$$0 \leq f'(x;x'-x) = \xi_{x'}^T(x'-x).$$

Then it follows that

$$f(x') \geq f(x) + \xi_{x'}^T(x'-x) \geq f(x),$$

which means that f attains its global minimum at x. $\qquad\square$

By recalling the theory of ε-subdifferentials we easily get the following optimality conditions.

Theorem 5.1.3. *If* $f : \mathbf{R}^n \to \mathbf{R}$ *is locally Lipschitz at* x *and attains its local minimum at* x, *then*

$$0 \in \partial_\varepsilon^G f(x). \tag{5.1}$$

Proof. This follows directly from Theorems 5.1.1 and 3.3.2.2. $\qquad\square$

Theorem 5.1.4. *If* $f : \mathbf{R}^n \to \mathbf{R}$ *is convex, then the following conditions are equivalent:*

(i) $0 \in \partial_\varepsilon f(x)$,

(ii) x *minimizes* f *within* ε, *i.e.* $f(x) \leq f(y) + \varepsilon$ *for all* $y \in \mathbf{R}^n$.

Proof. Suppose $0 \in \partial_\varepsilon f(x)$. By the definition of the ε-subdifferential this is equivalent to

$$f(y) \geq f(x) + 0^T(y-x) - \varepsilon = f(x) - \varepsilon \quad \text{for all} \quad y \in \mathbf{R}^n,$$

which means that x minimizes f within ε. $\qquad\square$

For constrained problems we need the following lemma.

Lemma 5.1.5. *Let f be Lipschitz with constant K on a set S. Let $x \in G \subset S$ and suppose that f attains a minimum over G at x. Then for any $\hat{K} \geq K$, the function $g(y) = f(y) + \hat{K} d_G(y)$ attains a minimum over S at x. If $\hat{K} > K$ and G is closed, then every minimizer of g over S must lie in G.*

Proof. Let us prove the first assertion by supposing the contrary. Then there would be a point $y \in S$ and $\varepsilon > 0$ such that $g(y) < g(x) - \hat{K}\varepsilon$; in other words

$$g(y) = f(y) + \hat{K} d_G(y) < f(x) + \hat{K} d_G(x) - \hat{K}\varepsilon = f(x) - \hat{K}\varepsilon.$$

Let $c \in G$ be such that $\|y - c\| \leq d_G(y) + \varepsilon$. Then by the Lipschitz condition

$$\begin{aligned}
f(c) &\leq f(y) + K\|y - c\| \leq f(y) + \hat{K}\|y - c\| \\
&\leq f(y) + \hat{K}(d_G(y + \varepsilon)) \\
&< f(x) + \hat{K}\varepsilon - \hat{K}\varepsilon = f(x),
\end{aligned}$$

which would contradict the fact that x minimizes f over G.

Suppose next that $\hat{K} > K$ and G is closed; suppose y also minimizes g over S. Then we can apply the first assertion to the constant $(\hat{K} + K)/2$ and obtain

$$f(y) + \hat{K} d_G(y) = f(x) = f(x) + (\hat{K} + K)d_G(x)/2 \leq f(y) + (\hat{K} + K)d_G(y)/2,$$

which implies that $\hat{K} d_G(y) \leq (\hat{K} + K)d_G(y)/2$. Since $\hat{K} > K$, this can be true only if $d_G(y) = 0$. Because G is closed, this implies, by Lemma 2.3.3, that $y \in G$. $\qquad\square$

Now it is easy to prove the corresponding conditions also for constrained problems.

Theorem 5.1.6. *If f is locally Lipschitz at x and attains its local minimum over the set $G \subset \mathbf{R}^n$ at x, then*

$$0 \in \partial f(x) + N_G(x). \tag{5.2}$$

Proof. Let $S \subset \mathbf{R}^n$ be an open set such that $x \in S$ and suppose f is Lipschitz with constant K on S. Then $G \cap S \subset S$ and f attains its minimum over $G \cap S$ at x. Then, by Lemma 5.1.5, the function $f(y) + K d_{G \cap S}(y)$ attains its minimum over S at x and by Theorem 5.1.1(i) we have

$$0 \in \partial(f + K d_{G \cap S})(x).$$

Now by applying Theorems 3.2.6, 4.1.7 and 4.1.9 we get

$$0 \in \partial(f + Kd_{G \cap S})(x) \subset \partial f(x) + K\partial d_{G \cap S}(x)$$
$$\subset \partial f(x) + N_{G \cap S}(x) = \partial f(x) + N_G(x).$$

$\square$

Theorem 5.1.7. *If $f : \mathbf{R}^n \to \mathbf{R}$ is convex and the set G is convex, then the following conditions are equivalent:*

(i) $0 \in \partial_c f(x) + N_G(x)$,

(ii) f *attains its global minimum over G at x.*

Proof. Theorem 5.1.6 implies that assertion (i) follows from (ii). Suppose now that $0 \in \partial_c f(x) + N_G(x)$. Then there exists $\xi \in \partial_c f(x)$ and $z \in N_G(x)$ such that $\xi = -z$. By the definition of the subdifferential for convex functions, we have

$$f(y) \geq f(x) + \xi^T(y - x) = f(x) - z^T(y - x) \qquad \text{for all} \quad y \in G$$

and, by Theorem 2.2.8, $z^T(y - x) \leq 0$, whence

$$f(y) \geq f(x) - z^T(y - x) \geq f(x) \qquad \text{for all} \quad y \in G.$$

$\square$

Quite often, in practice, the constraint set G has a special form. The following two corollaries generalize the classical Karush–Kuhn–Tucker optimality conditions.

Corollary 5.1.8. *Let $F : \mathbf{R}^n \to \mathbf{R}$ be a function such that $F(y) = \max\{F_i(y) \mid i = 1, \ldots, m\}$ where each $F_i : \mathbf{R}^n \to \mathbf{R}$ is locally Lipschitz at x and either $F(x) < 0$ or $0 \notin \partial F(x)$. Set*

$$G := \{y \in \mathbf{R}^n \mid F(y) \leq 0\}. \tag{5.3}$$

Suppose that $f : \mathbf{R}^n \to \mathbf{R}$ is locally Lipschitz at x and f attains its local minimum over the set G at x. Then there exist $\lambda_i \geq 0$ for $i = 1, \ldots, m$ such that $\lambda_i F_i(x) = 0$ and

$$0 \in \partial f(x) + \sum_{i=1}^{m} \lambda_i \partial F_i(x). \tag{5.4}$$

Proof. The function F is evidently locally Lipschitz at x. If $F(x) < 0$, then $x \in \text{int } G$ and, by Theorem 4.1.8, $N_G(x) = \{0\}$. Then it follows from Theorem 5.1.6 that $0 \in \partial f(x)$ so we can take $\lambda_i := 0$ for each $i = 1, \ldots, m$.

Suppose now that $F(x) = 0$, so the set $I(x) := \{i \in \{1, \ldots, m\} \mid F_i(x) = F(x)\}$ is nonempty. Then we can write the set G in the form

$$G = \{y \in \mathbf{R}^n \mid (y, F(x)) \in \text{epi } F\}.$$

The fact that $0 \notin \partial F(x)$ implies, by Theorem 4.2.5, that

$$N_G(x) \subset \bigcup_{\lambda \geq 0} \lambda \partial F(x).$$

On the other hand by Theorem 3.2.13 we have

$$\partial F(x) \subset \text{conv}\{\partial F_i(x) \mid i \in I(x)\},$$

thus by Theorem 5.1.6 there exist $\lambda_i \geq 0$ such that $\lambda_i F_i(x) = 0$ and

$$0 \in \partial f(x) + \sum_{i=1}^{m} \lambda_i \partial F_i(x).$$

$\square$

Corollary 5.1.9. *Let $F : \mathbf{R}^n \to \mathbf{R}$ be a function such that $F(y) = \max\{F_i(y) \mid i = 1, \ldots, m\}$ where each $F_i : \mathbf{R}^n \to \mathbf{R}$ is convex and $F(z) < 0$ for some $z \in \mathbf{R}^n$. Set*

$$G := \{y \in \mathbf{R}^n \mid F(y) \leq 0\} \tag{5.5}$$

Suppose $f : \mathbf{R}^n \to \mathbf{R}$ is convex. Then the following conditions are equivalent:

(i) *f attains its global minimum over G at x,*

(ii) *there exist $\lambda_i \geq 0$ for $i = 1, \ldots, m$ such that $\lambda_i F_i(x) = 0$ and*

$$0 \in \partial_c f(x) + \sum_{i=1}^{m} \lambda_i \partial_c F_i(x). \tag{5.6}$$

Proof. It is evident that F is convex, being a maximum of convex functions. Because there exists $z \in \mathbf{R}^n$ such that $F(z) < 0$, we see that if $F(x) = 0$, then x is not a global minimum of F over G. Then by Theorem 5.1.2 we have $0 \notin \partial_c F(x)$ and Corollary 5.1.8 implies that assertion (ii) follows from (i).

Suppose now that (ii) holds. Then

$$\sum_{i=1}^{m} \lambda_i \partial_c F_i(x) \subset \text{conv}\{\partial_c F_i(x) \mid i \in I(x)\}.$$

By Theorem 3.2.2(b) the function F is regular, so by Theorem 3.2.13

$$\text{conv}\,\{\partial_c F_i(x) \mid i \in I(x)\} = \partial_c F(x)$$

and by Theorem 4.2.5 one has

$$\partial_c F(x) \subset \bigcup_{\lambda \geq 0} \lambda \partial_c F(x) = N_G(x).$$

Thus we have

$$0 \in \partial_c f(x) + N_G(x),$$

whence by Theorem 5.1.7 f attains its global minimum over G at x.   □

5.2. Linearization of Unconstrained Problem

In this section we define some notions of linearization for Lipschitz functions and present their basic properties. With these linearizations we can construct a piecewise linear local approximation to the unconstrained optimization problem. This approximation will be used in nonsmooth optimization methods in the next part.

Definition 5.2.1. Suppose the function $f : \mathbf{R}^n \to \mathbf{R}$ is locally Lipschitz at $x \in \mathbf{R}^n$ and let $\xi \in \partial f(x)$ be an arbitrary subgradient. Then *the ξ-linearization of f at x* is the function $\bar{f}_\xi : \mathbf{R}^n \to \mathbf{R}$ defined by

$$\bar{f}_\xi(y) := f(x) + \xi^{\mathrm{T}}(y - x) \qquad \text{for all} \quad y \in \mathbf{R}^n \tag{5.7}$$

and *the linearization of f at x* is the function $\hat{f}_x : \mathbf{R}^n \to \mathbf{R}$ such that

$$\hat{f}_x(y) := \max\,\{\bar{f}_\xi(y) \mid \xi \in \partial f(x)\} \qquad \text{for all} \quad y \in \mathbf{R}^n. \tag{5.8}$$

Some elementary facts about these linearizations will be noted.

Theorem 5.2.2. *Let $f : \mathbf{R}^n \to \mathbf{R}$ be locally Lipschitz at x. Then the linearization $\hat{f}_x$ is convex and*

(i) $\hat{f}_x(x) = f(x)$,

(ii) $\hat{f}_x(y) = f(x) + f^\circ(x; y - x)$ *for all* $y \in \mathbf{R}^n$,

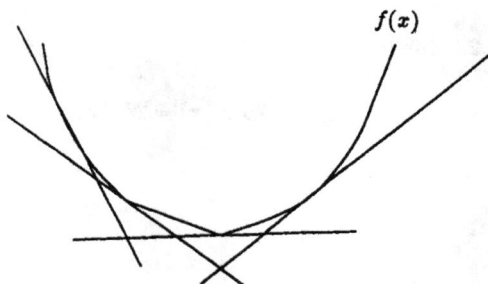

$f(x)$

Figure 5.1. The linearizations

(iii) $\partial_c \hat{f}_x(x) = \partial f(x)$.

Proof. We show first that $\hat{f}_x$ is convex. To see this, let $y, z \in \mathbf{R}^n$ and $\lambda \in [0, 1]$. Then

$$
\begin{aligned}
\hat{f}_x(\lambda y + (1 - \lambda)z) &= \max \left\{ f(x) + \xi^T (\lambda y + (1 - \lambda)z) \mid \xi \in \partial f(x) \right\} \\
&\leq \lambda \cdot \max \left\{ f(x) + \xi^T y \mid \xi \in \partial f(x) \right\} \\
&\quad + (1 - \lambda) \cdot \max \left\{ f(x) + \xi^T z \mid \xi \in \partial f(x) \right\} \\
&= \lambda \hat{f}_x(y) + (1 - \lambda) \hat{f}_x(z).
\end{aligned}
$$

By (5.7) and (5.8) we have

$$
\hat{f}_x(x) = \max \left\{ f(x) + \xi^T (x - x) \mid \xi \in \partial f(x) \right\} = f(x),
$$

which establishes (i). Assertion (ii) follows from Theorem 3.1.4(ii) because

$$
\begin{aligned}
\hat{f}_x(y) &= \max \{ f(x) + \xi^T (y - x) \mid \xi \in \partial f(x) \} \\
&= f(x) + \max_{\xi \in \partial f(x)} \xi^T (y - x) \\
&= f(x) + f^\circ(x; y - x) \qquad \text{for all} \quad y \in \mathbf{R}^n.
\end{aligned}
$$

For (iii) let $\xi \in \partial f(x)$. Then by assertion (i) we have

$$
\hat{f}_x(y) \geq \bar{f}_\xi(y) = f(x) + \xi^T (y - x) = \hat{f}_x(x) + \xi^T (y - x),
$$

so $\xi \in \partial_c \hat{f}_x(x) = \partial \hat{f}_x(x)$. On the other hand, by assertion (ii) we have

$$\hat{f}_x^{\circ}(x;v) = \limsup_{\substack{y \to x \\ \lambda \downarrow 0}} \frac{\hat{f}_x(y+\lambda v) - \hat{f}_x(y)}{\lambda}$$

$$= \limsup_{\substack{y \to x \\ \lambda \downarrow 0}} \frac{f(x) + f^{\circ}(x; y + \lambda v - x) - f(x) - f^{\circ}(x; y - x)}{\lambda}$$

and by Theorem 3.1.2(i)

$$\hat{f}_x^{\circ}(x;v) \le \limsup_{\substack{y \to x \\ \lambda \downarrow 0}} \frac{f^{\circ}(x; y - x) + f^{\circ}(x; \lambda v) - f^{\circ}(x; y - x)}{\lambda}$$

$$= \limsup_{\substack{y \to x \\ \lambda \downarrow 0}} \frac{\lambda f^{\circ}(x;v)}{\lambda}$$

$$= f^{\circ}(x;v),$$

so $\partial_c \hat{f}_x(x) \subset \partial f(x)$ and the proof is complete. $\quad\square$

If the function f happens to be convex, then we get the following useful result.

Theorem 5.2.3. *Let* $f : \mathbf{R}^n \to \mathbf{R}$ *be convex. Then*

(i) $f(y) = \max \{ \hat{f}_x(y) \mid x \in \mathbf{R}^n \}$ *for all* $y \in \mathbf{R}^n$,

(ii) $\hat{f}_x(y) \le f(y)$ *for all* $y \in \mathbf{R}^n$,

(iii) epi $f \subset$ epi $\hat{f}_x$.

Proof. The convexity of f implies by Theorem 2.1.7 that for all $y \in \mathbf{R}^n$ we have

$$f(y) = \max \{ f(x) + \xi^T(y - x) \mid \xi \in \partial f(x), \ x \in \mathbf{R}^n \}$$
$$= \max \{ \bar{f}_\xi(y) \mid \xi \in \partial f(x), \ x \in \mathbf{R}^n \}$$
$$= \max \{ \hat{f}_x(y) \mid x \in \mathbf{R}^n \}.$$

The assertions (ii) and (iii) now follow directly from (i). $\quad\square$

The fundamental problem in iterative optimization methods is to find a direction such that the function values will decrease on moving in this direction. For this reason we shall need the next concept.

Definition 5.2.4. The direction $d \in \mathbf{R}^n$ is said to be a *descent direction* for $f : \mathbf{R}^n \to \mathbf{R}$ at x if there exists $\varepsilon > 0$ such that

$$f(x + td) < f(x) \qquad \text{for all} \quad t \in (0, \varepsilon]. \tag{5.9}$$

The next theorem indicates how to find descent directions for a Lipschitz function.

Theorem 5.2.5. Let $f : \mathbf{R}^n \to \mathbf{R}$ be locally Lipschitz at x. The direction $d \in \mathbf{R}^n$ is a descent direction for f at x if any of the following hold:

 (i) $f^\circ(x; d) < 0$,

 (ii) $\xi^T d < 0$ for all $\xi \in \partial f(x)$,

 (iii) $\xi^T d < 0$ for all $\xi \in \partial_\varepsilon^G f(x)$,

 (iv) d is a descent direction for $\hat{f}_x$ at x.

Proof. By the definition of the generalized directional derivative,

$$\limsup_{t \downarrow 0} \frac{f(x + td) - f(x)}{t} \le \limsup_{\substack{y \to x \\ t \downarrow 0}} \frac{f(y + td) - f(y)}{t} = f^\circ(x; d) < 0.$$

Then, by the definition of upper limit, there exists $\varepsilon > 0$ such that $f(x + td) - f(x) < 0$ for all $t \in (0, \varepsilon]$, which means that d is a descent direction for f at x. Part (ii) follows from (i), since Theorem 3.1.4(ii) gives

$$f^\circ(x; d) = \max\{\xi^T d \mid \xi \in \partial f(x)\} < 0$$

and part (iii) follows from (ii), since Theorem 3.3.2.2 gives

$$\partial f(x) \subset \partial_\varepsilon^G f(x).$$

In order to prove assertion (iv), let d be a descent direction for $\hat{f}_x$ at x. Then there exists $\varepsilon > 0$ such that

$$\hat{f}_x(x + td) < \hat{f}_x(x) \qquad \text{for all} \quad t \in (0, \varepsilon].$$

Then by Theorems 5.2.2(i), (ii) and 3.1.2(i) we have

$$0 > f(x) + f^\circ(x; x + td - x) - f(x) = t \cdot f^\circ(x; d) \qquad \text{for all} \quad t \in (0, \varepsilon],$$

so $f^\circ(x; d) < 0$ and then, by part (i), d is a descent direction for f at x. $\qquad \square$

For the main result of this section we need the following two lemmas.

Lemma 5.2.6. *Let $G \subset \mathbf{R}^n$ be a convex compact set. Then for $p \in G$*

$$p = \arg\min\{\|g\| \mid g \in G\} \iff p^{\mathrm{T}}g \geq \|p\|^2 \quad \text{for all} \quad g \in G. \quad (5.10)$$

Proof. We show first that $\arg\min\{\|g\| \mid g \in G\}$ is well-defined. The function $\|\cdot\|$ is evidently continuous, so it attains its minimum over the compact set G. If $0 \in G$, then it is clear that zero is a unique minimizer. To see that the minimum is unique also in the case when $0 \notin G$, suppose that there exist two minimum points p_1 and p_2, with $p_1 \neq p_2 \neq 0$. Let $\lambda \in (0,1)$; then by convexity of G the point $\lambda p_1 + (1-\lambda)p_2$ is in G. If p_1 and p_2 are linearly independent, we have

$$\|\lambda p_1 + (1-\lambda)p_2\| < \lambda\|p_1\| + (1-\lambda)\|p_2\| = \arg\min\{\|g\| \mid g \in G\},$$

which is impossible. On the other hand if p_1 and p_2 are linearly dependent, then there exists $\mu \neq 1$ such that $p_2 = \mu p_1$. If $\mu \leq 0$, it follows that $0 \in G$, which is impossible and if $\mu > 0$, it follows that

$$\|p_2\| = \mu\|p_1\| \neq \|p_1\|,$$

which is impossible because both p_1 and p_2 were minimum points. Thus the minimum is unique and $\arg\min\{\|g\| \mid g \in G\}$ is well-defined.

Suppose first that $p^{\mathrm{T}}g \geq \|p\|^2$ for all $g \in G$. Then we have

$$\|p\|^2 \leq p^{\mathrm{T}}g \leq \|p\|\|g\| \quad \text{for all} \quad g \in G,$$

whence

$$\|p\| \leq \arg\min\{\|g\| \mid g \in G\}$$

and so $p = \arg\min\{\|g\| \mid g \in G\}$. Suppose next that $p = \arg\min\{\|g\| \mid g \in G\}$ and there exists $g \in G$ such that $g^{\mathrm{T}}p < \|p\|^2$. Because G is convex the point $\lambda g + (1-\lambda)p$ is in G for each $\lambda \in (0,1)$ and

$$\begin{aligned}
\|\lambda g + (1-\lambda)p\|^2 &= \|p + \lambda(g-p)\|^2 \\
&= (p + \lambda(g-p))^{\mathrm{T}}(p + \lambda(g-p)) \\
&= \|p\|^2 + 2\lambda[g^{\mathrm{T}}p - \|p\|^2] + \lambda^2\|g-p\|^2 \\
&< \|p\|^2,
\end{aligned}$$

when λ is small enough. This would contradict the fact that p is the minimum point so $g^{\mathrm{T}}p \geq \|p\|^2$ for all $g \in G$, which completes the proof. $\quad\square$

Lemma 5.2.7. *Suppose* $f : \mathbf{R}^n \to \mathbf{R}$ *is locally Lipschitz at* x *and* $d \in \mathbf{R}^n$ *is an arbitrary direction. Then the function* $f^\circ(x; \cdot)$ *is locally Lipschitz at* d *and*

$$\partial f^\circ(x; d) \subset \{\xi \in \partial f(x) \mid \xi^T d = f^\circ(x; d)\}. \tag{5.11}$$

Proof. By Theorem 3.1.2(ii) the function $f^\circ(x; \cdot)$ is locally Lipschitz at d. Let us suppose that $\xi \in \partial f^\circ(x; d)$. Then by Theorem 3.1.2(i) we have for all $v \in \mathbf{R}^n$

$$\begin{aligned}
\xi^T v \leq (f^\circ)^\circ((x; d); v) &= \limsup_{\substack{d' \to d \\ \lambda \downarrow 0}} \frac{f^\circ(x; d' + \lambda v) - f^\circ(x; d')}{\lambda} \\
&\leq \limsup_{\substack{d' \to d \\ \lambda \downarrow 0}} \frac{f^\circ(x; d') + \lambda f^\circ(x; v) - f^\circ(x; d')}{\lambda} = \limsup_{\substack{d' \to d \\ \lambda \downarrow 0}} f^\circ(x; v) \\
&= f^\circ(x; v),
\end{aligned}$$

so $\xi \in \partial f(x)$. Now we have

$$\begin{aligned}
f^\circ(x; d) \geq (f^\circ)^\circ((x; d); d) \\
&= \limsup_{\substack{d' \to d \\ \lambda \downarrow 0}} \frac{f^\circ(x; d' + \lambda d) - f^\circ(x; d')}{\lambda} \\
&\geq \limsup_{\substack{d' \to d \\ \lambda \downarrow 0}} \frac{f^\circ(x; (1 + \lambda)d) - f^\circ(x; d)}{\lambda} \\
&= \limsup_{\substack{d' \to d \\ \lambda \downarrow 0}} \frac{f^\circ(x; d) + \lambda f^\circ(x; d) - f^\circ(x; d)}{\lambda} \\
&= f^\circ(x; d),
\end{aligned}$$

so $f^\circ(x; d) = (f^\circ)^\circ((x; d); d)$. If one had $\xi^T d < f^\circ(x; d)$, then

$$-(f^\circ)^\circ((x; d); -d) \leq -\xi^T(-d) = \xi^T d < f^\circ(x; d) = (f^\circ)^\circ((x; d); d)$$

and thus

$$\begin{aligned}
\liminf_{\substack{d' \to d \\ \lambda \downarrow 0}} \frac{f^\circ(x; d' + \lambda d) - f^\circ(x; d')}{\lambda} &= -\limsup_{\substack{d' \to d \\ \lambda \downarrow 0}} \frac{f^\circ(x; d') - f^\circ(x; d' + \lambda d)}{\lambda} \\
&= -\limsup_{\substack{d' \to d \\ \lambda \downarrow 0}} \frac{f^\circ(x; d' + \lambda d - \lambda d) - f^\circ(x; d' + \lambda d)}{\lambda} = -(f^\circ)^\circ((x; d); -d) \\
&< (f^\circ)^\circ((x; d); d) = \limsup_{\substack{d' \to d \\ \lambda \downarrow 0}} \frac{f^\circ(x; d' + \lambda d) - f^\circ(x; d')}{\lambda},
\end{aligned}$$

so

$$\liminf_{\substack{d' \to d \\ \lambda \downarrow 0}} \frac{f^\circ(x; d' + \lambda d) - f^\circ(x; d')}{\lambda} < \limsup_{\substack{d' \to d \\ \lambda \downarrow 0}} \frac{f^\circ(x; d' + \lambda d) - f^\circ(x; d')}{\lambda}.$$

This means that there exists no directional derivative $(f^\circ)'(x; d)$. On the other hand by Theorem 3.1.2(i) the function $v \mapsto f^\circ(x; v)$ is positively homogeneous and subadditive, so it is convex and by Theorem 3.2.2(b) the convex function is regular which would contradict the fact that there exists no directional derivative $(f^\circ)'(x; d)$. Thus we have

$$\xi^{\mathrm{T}} d = f^\circ(x; d)$$

and the proof is complete. □

The most promising nonsmooth optimization methods are based on the following theorem. It tells how one can find a descent direction for the linearization function. Note that by Theorem 5.2.5(iv) this direction is a descent direction also for the original function.

Theorem 5.2.8. *Suppose that* $f : \mathbf{R}^n \to \mathbf{R}$ *is locally Lipschitz at* x *and* $\xi^* \in \partial f(x)$ *is such that* $\xi^* = \arg\min\{\|\xi\| \mid \xi \in \partial f(x)\}$. *Consider the problem*

$$\text{minimize} \quad \hat{f}_x(x + d) + \tfrac{1}{2}\|d\|^2 \quad \text{over all} \quad d \in \mathbf{R}^n. \tag{5.12}$$

Then

(i) *The problem* (5.12) *has a unique solution* $d^* \in \mathbf{R}^n$ *such that* $-d^* = \xi^* \in \partial f(x)$,

(ii) $f^\circ(x; d^*) = -\|d^*\|^2$,

(iii) $\hat{f}_x(x + \lambda d^*) = \hat{f}_x(x) - \lambda\|\xi^*\|^2 \quad$ *for all* $\lambda \in [0, 1]$,

(iv) $0 \notin \partial f(x) \iff d^* \neq 0$,

(v) $0 \in \partial f(x) \iff \hat{f}_x$ *attains its global minimum at* x.

Proof. We show first assertions (i) and (ii). Define a function $\rho : \mathbf{R}^n \to \mathbf{R}$ such that for all $d \in \mathbf{R}^n$

$$\rho(d) := \hat{f}_x(x + d) + \tfrac{1}{2}\|d\|^2 = f(x) + f^\circ(x; d) + \tfrac{1}{2}\|d\|^2.$$

To see that the function ρ is strictly convex, let $d, d' \in \mathbf{R}^n$ and $\lambda \in (0,1)$. Then by Theorem 3.1.2(i) and the fact that the mapping $t \mapsto t^2$ is strictly convex we get

$$
\begin{aligned}
\rho(\lambda d + (1-\lambda)d') &= f(x) + f^\circ(x; \lambda d + (1-\lambda)d') + \tfrac{1}{2}\|\lambda d + (1-\lambda)d'\|^2 \\
&< f(x) + \lambda f^\circ(x;d) + (1-\lambda)f^\circ(x;d') \\
&\quad + \tfrac{1}{2}\lambda^2\|d\|^2 + \tfrac{1}{2}(1-\lambda)^2\|d'\|^2 \\
&< \lambda(f(x) + f^\circ(x;d) + \tfrac{1}{2}\|d\|^2) \\
&\quad + (1-\lambda)(f(x) + f^\circ(x;d') + \tfrac{1}{2}\|d'\|^2) \\
&= \lambda\rho(d) + (1-\lambda)\rho(d').
\end{aligned}
$$

Since the function ρ is strictly convex and for all $\xi \in \partial f(x)$ one has

$$
\rho(d) \geq f(x) + \xi^T d + \tfrac{1}{2}\|d\|^2 \geq f(x) - \|\xi\|\|d\| + \tfrac{1}{2}\|d\|^2 \longrightarrow \infty,
$$

as $\|d\| \to \infty$, there exists a unique $d^* \in \mathbf{R}^n$ which minimizes ρ. Then by Theorem 5.1.2 we have $0 \in \partial\rho(d^*)$. The fact that the functions $d \mapsto f^\circ(x;d)$ and $d \mapsto \tfrac{1}{2}\|d\|^2$ are convex and hence regular implies that

$$
\partial\rho(d^*) = \partial(f^\circ(x;d^*) + \tfrac{1}{2}\|d^*\|^2) = \partial f^\circ(x;d^*) + \partial(\tfrac{1}{2}\|d^*\|^2).
$$

The function $d \mapsto \tfrac{1}{2}\|d\|^2$ is also continuously differentiable, whence, by Theorem 3.1.7,

$$
\partial(\tfrac{1}{2}\|d^*\|^2) = d^*
$$

and Lemma 5.2.7 implies that

$$
\partial f^\circ(x;d^*) \subset \{\xi \in \partial f(x) \mid \xi^T d^* = f^\circ(x;d^*)\}.
$$

So we have

$$
0 \in d^* + \{\xi \in \partial f(x) \mid \xi^T d^* = f^\circ(x;d^*)\},
$$

which means that there exists $\hat{\xi} \in \partial f(x)$ such that $\hat{\xi} = -d^*$ and $f^\circ(x;d^*) = -\|\hat{\xi}\|^2$. The fact that $\hat{\xi} \in \partial f(x)$ and that

$$
\|\hat{\xi}\|^2 = -f^\circ(x;d^*) \leq -\xi^T d^* = \xi^T \hat{\xi},
$$

for all $\xi \in \partial f(x)$ implies, by Lemma 5.2.6, that $\hat{\xi} = \arg\min\{\|\xi\| \mid \xi \in \partial f(x)\} = \xi^*$, which establishes assertions (i) and (ii). Assertion (iii) follows from Theorems 5.2.2, 3.1.2(i) and assertion (ii) of this theorem because

$$
\hat{f}_x(x + \lambda d^*) = f(x) + f^\circ(x; \lambda d^*) = \hat{f}_x(x) + \lambda f^\circ(x; d^*) = \hat{f}_x(x) - \lambda\|\xi^*\|^2.
$$

To see assertion (iv) suppose that $0 \notin \partial f(x)$. This is equivalent to

$$\|d^*\| = \|\xi^*\| = \min_{\xi \in \partial f(x)} \|\xi\| > 0,$$

which is equivalent to $d^* \neq 0$. Assertion (v) follows directly from Theorems 5.2.2 and 5.1.2(ii). □

5.3. Linearization of Constrained Problem

In this section we consider constrained optimization. As is well known from classical nonlinear optimization theory the constraints will always cause some difficulties. For this reason we shall need some new concepts.

Consider the following constrained optimization problem:

$$\text{minimize} \quad f(y) \quad \text{subject to} \quad y \in G, \tag{5.13}$$

where the *objective function* $f : \mathbf{R}^n \to \mathbf{R}$ is supposed to be Lipschitz on the nonempty *feasible set* $G := G_F \cap G_C$ defined by

$$\begin{aligned} G_F &:= \{x \in \mathbf{R}^n \mid F(x) \leq 0\} \quad \text{and} \\ G_C &:= \{x \in \mathbf{R}^n \mid Cx \leq b\}. \end{aligned} \tag{5.14}$$

Here C is an $m_C \times n$ constrain matrix, C_i for $i \in I = \{1, \dots, m_C\}$ are the rows of C, the right-hand side b is an m_C-vector.

Definition 5.3.1. The *total constraint function* $F : \mathbf{R}^n \to \mathbf{R}$ is defined by

$$F(x) := \max\{F_i(x) \mid i = 1, \dots, m_F\} \quad \text{for all} \quad x \in \mathbf{R}^n, \tag{5.15}$$

where the *constraint functions* $F_i : \mathbf{R}^n \to \mathbf{R}$ for $i = 1, \dots, m_F$ are supposed to be Lipschitz functions. The function $H(\cdot\,; x) : \mathbf{R}^n \to \mathbf{R}$ such that

$$H(y; x) := \max\{f(y) - f(x),\, F(y)\} \quad \text{for all} \quad y \in \mathbf{R}^n \tag{5.16}$$

is called the *improvement function* at x. The problem is said to be convex if the functions f and F_i are convex.

Our aim is to handle the nonsmooth constraints by transforming the generally constrained problem to only a constrained problem by minimizing $H(\cdot\,; x) : \mathbf{R}^n \to \mathbf{R}$ subject to the linear constraints. The following shows how to compute the subdifferential of the improvement function.

Theorem 5.3.2. *If* $x \in \mathbf{R}^n$, *then*

$$\partial H(y;x) \subset \begin{cases} \partial f(y) & \text{if } f(y) > F(y) + f(x), \\ \text{conv}\{\partial f(y) \cup \partial F(y)\} & \text{if } f(y) = F(y) + f(x), \\ \partial F(y) & \text{if } f(y) < F(y) + f(x). \end{cases} \tag{5.17}$$

If, in addition, the problem functions f *and* F_i *for* $i = 1, \ldots, m$ *are regular at* y, *then equality holds in* (5.17).

Proof. This follows directly from Theorem 3.2.13.   □

For convenience we define the following constraint qualification.

Definition 5.3.3. The problem (5.13) is said to satisfy the *Cottle constraint qualification* at x if either $F(x) < 0$ or there do not exist $\nu_i \geq 0$ for $i \in I$ such that $\nu_i[C_i^T x - b_i] = 0$ and

$$0 \in \partial F(x) + \sum_{i \in I} \nu_i C_i. \tag{5.18}$$

Furthermore, the problem (5.13) is said to satisfy the *Slater constraint qualification* if

$$F(z) < 0 \quad \text{for some} \quad z \in G. \tag{5.19}$$

Theorem 5.3.4. *If the constraint functions* F_i *for* $i = 1, \ldots, m_F$ *are convex, then the Cottle constraint qualification at* $x \in G$ *is equivalent to the Slater constraint qualification.*

Proof. Assume the Slater constraint qualification. If $F(x) = 0$, then the fact that there exists $z \in G_C$ such that $F(z) < 0 = F(x)$ is equivalent to x cannot be a global minimum of F over G_C. From Theorem 5.1.7 we derive

$$0 \notin \partial F(x) + N_{G_C}.$$

By Corollary 5.1.9 we have

$$N_{G_C}(x) = \bigcup_{\nu \geq 0} \nu\{C_i \mid C_i^T x = b_i\},$$

thus there do not exist $\nu_i \geq 0$ for $i \in I$ such that $\nu_i[C_i^T x - b_i] = 0$ and

$$0 \in \partial F(x) + \sum_{i \in I} \nu_i C_i.$$

                                                                        □

Now we get the following optimality conditions.

Corollary 5.3.5. *Suppose the problem* (5.13) *satisfies the Cottle constraint qualification at* $x \in G$. *Then we have the following necessary conditions for* x *to be a local solution of* (5.13).

(i) $H(\cdot; x)$ *attains a local minimum over* G_C *at* x,

(ii) *there exist* $\nu_i \geq 0$ *for* $i \in I$ *such that* $\nu_i[C_i^T x - b_i] = 0$ *and*

$$0 \in \partial H(x; x) + \sum_{i \in I} \nu_i C_i, \tag{5.20}$$

(iii) *there exist* $\mu_i \geq 0$ *for* $i = 1, \ldots, m_F$ *and* $\nu_i \geq 0$ *for* $i \in I$ *such that* $\mu_i F_i(x) = 0$, $\nu_i[C_i^T x - b_i] = 0$ *and*

$$0 \in \partial f(x) + \sum_{i=1}^{m_F} \mu_i \partial F_i(x) + \sum_{i \in I} \nu_i C_i. \tag{5.21}$$

If the functions f *and* F_i *for* $i = 1, \ldots, m_F$ *are convex and the problem* (5.13) *satisfies the Slater constraint qualification, then each of the conditions* (i)-(iii) *is also sufficient and the solution is global.*

Proof. Suppose that the problem (5.13) has a local solution at $x \in G$. Then there exists $\varepsilon > 0$ such that $f(x) \leq f(y)$ for all $y \in B(x; \varepsilon) \cap G$ and because $x \in G_F$ we have $H(x; x) = 0$. If there were $y_0 \in B(x; \varepsilon) \cap G$ such that $H(y_0; x) < 0$, then $F(y_0) < 0$ and $f(y_0) < f(x)$. This would mean that $y_0 \in G$ and $f(y_0) < f(x)$, which is impossible. Thus $H(\cdot; x)$ attains a local minimum over G_C at x. Assertions (ii) and (iii) follow from (i) by the proof of Theorem 5.3.4 and Theorem 5.1.8, respectively.

Suppose next that the functions f and F_i for $i = 1, \ldots, m_F$ are convex and the problem (5.13) satisfies the Slater constraint qualification. Then by Theorem 5.3.2 $H(\cdot; x)$ is convex and then the condition (5.20), by Theorem 5.1.7, is also sufficient for $H(\cdot; x)$ to have a global minimum over G_C at x. Above we saw that (ii) implies (iii). To see the reverse, note that by convexity, Theorem 5.3.2 implies that

$$[\partial f(x) + \sum_{i=1}^{m_F} \mu_i \partial F_i(x)] / (1 + \sum_{i=1}^{m_F} \mu_i) \subset \text{conv} \{\partial f(x) \cup \partial F(x)\} = \partial H(x; x).$$

Now to complete the proof we have to show that condition (i) is equivalent with f attaining its global minimum over G at x. Suppose that x is a global minimum for f over G. Then $f(x) \leq f(y)$ for all $y \in G$ and $H(x; x) = 0$. If there were $y_0 \in G_C$ such that $H(y_0; x) < 0$, then $f(y_0) < f(x)$ and $F(y_0) < 0$. This would mean that $f(y_0) < f(x)$ and $y_0 \in G$, which is impossible and so we have (i).

Next, suppose that condition (i) holds, but $f(y_0) < f(x)$ for some $y_0 \in G$. By the Slater constraint qualification there exists $z \in G_C$ such that $F(z) < 0$. Then by convexity we can find $\varepsilon > 0$ such that for all $\lambda \in (0, \varepsilon)$

$$f(\lambda z + (1 - \lambda)y_0) \leq \lambda f(z) + (1 - \lambda)f(y_0) < f(x).$$

Then

$$F(\lambda z + (1 - \lambda)y_0) \leq \lambda F(z) + (1 - \lambda)F(y_0) \leq \lambda F(z) < 0$$

and

$$C(\lambda z + (1 - \lambda)y_0) = \lambda C z + (1 - \lambda)C y_0 \leq b,$$

from which it follows that $y_0 \in G_C$ and

$$H(\lambda z + (1 - \lambda)y_0; x) < 0 = H(x; x),$$

which contradicts assertion (i). Thus (i) is also a sufficient condition and the proof is complete. $\qquad\square$

As in the unconstrained case, we define the following linearizations.

Definition 5.3.6. The ξ-*linearizations* of functions f and F_i for $i = 1, \ldots, m_F$ at $x \in G$ are defined for all $y \in \mathbf{R}^n$ by

$$\begin{cases} \bar{f}_\xi(y) := f(x) + \xi^{\mathrm{T}}(y - x) & \text{for} \quad \xi \in \partial f(x) \\ \bar{F}_{i,\xi}(y) := F_i(x) + \xi^{\mathrm{T}}(y - x) & \text{for} \quad \xi \in \partial F_i(x) \end{cases} \tag{5.22}$$

and the *linearizations* of f and F_i at x are defined for all $y \in \mathbf{R}^n$ by

$$\begin{cases} \hat{f}_x(y) := \max\{\bar{f}_\xi(y) \mid \xi \in \partial f(x)\} \\ \hat{F}_{i,x}(y) := \max\{\bar{F}_{i,\xi}(y) \mid \xi \in \partial F_i(x)\}. \end{cases} \tag{5.23}$$

Then the *linearizations* of F and $H(\cdot; x)$ at x are defined for all $y \in \mathbf{R}^n$ by

$$\begin{cases} \hat{F}_x(y) := \max\{\hat{F}_{i,x}(y) \mid i = 1, \ldots, m_F\} \\ \hat{H}_x(y) := \max\{\hat{f}_x(y) - \hat{f}_x(x), \hat{F}_x(y)\}. \end{cases} \tag{5.24}$$

Some elementary facts about these linearizations will now be noted.

Theorem 5.3.7. *The linearizations $\hat{f}_x$, $\hat{F}_x$ and $\hat{H}_x$ are convex such that $\hat{f}_x(x) = f(x)$, $\hat{F}_x(x) = F(x) \leq 0$, $\hat{H}_x(x) = H(x; x) = 0$ and*

$$\begin{cases} \partial f(x) = \partial \hat{f}_x(x) \\ \partial F(x) \subset \partial \hat{F}_x(x) = \text{conv}\{\partial F_i(x) \mid i \in I(x)\} \\ \partial H(x; x) \subset \partial \hat{H}_x(x) = \begin{cases} \partial f(x) & \text{if } F(x) < 0, \\ \text{conv}\{\partial f(x) \cup \partial \hat{F}_x(x)\} & \text{if } F(x) = 0. \end{cases} \end{cases} \quad (5.25)$$

If in addition the problem functions f and F_i for $i = 1, \ldots, m_F$ are convex then equality holds in (5.25) and for all $y \in \mathbf{R}^n$ we have

$$\begin{cases} \hat{f}_x(y) \leq f(y) = \max\{\hat{f}_x(y) \mid x \in \mathbf{R}^n\} \\ \hat{F}_x(y) \leq F(y) \\ \hat{H}_x(y) \leq H(y; x). \end{cases} \quad (5.26)$$

Proof. Follows from Theorems 5.2.2, 5.3.2 and 3.2.13.   □

As a result of previous theorem we get the same optimality conditions as for the improvement function.

Corollary 5.3.8. *Suppose the problem (5.13) satisfies the Cottle constraint qualification at $x \in G$. Then we derive the following necessary conditions for x to be a local solution of (5.13).*

(i) *$\hat{H}_x$ attains a global minimum over G_C at x,*

(ii) *there exist $\nu_i \geq 0$ for $i \in I$ such that $\nu_i[C_i^\mathrm{T} x - b_i] = 0$ and*

$$0 \in \partial \hat{H}_x(x) + \sum_{i \in I} \nu_i C_i, \quad (5.27)$$

(iii) *there exist $\mu_i \geq 0$ for $i = 1, \ldots, m_F$ and $\nu_i \geq 0$ for $i \in I$ such that $\mu_i F_i(x) = 0$, $\nu_i[C_i^\mathrm{T} x - b_i] = 0$ and*

$$0 \in \partial f(x) + \sum_{i=1}^{m_F} \mu_i \partial F_i(x) + \sum_{i \in I} \nu_i C_i. \quad (5.28)$$

If the functions f and F_i for $i = 1, \ldots, m_F$ are convex and the problem (5.13) satisfies the Slater constraint qualification, then each of the conditions (i)-(iii) is also sufficient and the solution is global.

Proof. Follows directly from Theorems 5.3.5 and 5.3.7.   □

In constrained optimization it is not enough to find any descent direction, since we are not allowed to violate the constraints. For this reason we define the following new concept.

Definition 5.3.9. The direction $d \in \mathbf{R}^n$ is called a *feasible direction* subject to G at x, if there exists $\delta > 0$ such that

$$x + td \in G \qquad \text{for all} \quad t \in (0, \delta]. \tag{5.29}$$

The next theorem tells how one can find a feasible descent direction.

Theorem 5.3.10. *If the direction $d \in \mathbf{R}^n$ is a descent direction for $\hat{H}_x$ at $x \in G$ and feasible subject to G_C, then d is a descent direction for f at x and feasible subject to G.*

Proof. Suppose $d \in \mathbf{R}^n$ is a descent direction for $\hat{H}_x$ at $x \in G$ and feasible subject to G_C. Then there exists $\varepsilon > 0$ such that

$$\hat{H}_x(x + td) < \hat{H}_x(x) = 0 \qquad \text{for all} \quad t \in (0, \varepsilon] \tag{5.30}$$

and $\delta' > 0$ such that

$$x + td \in G_C \qquad \text{for all} \quad t \in (0, \delta']. \tag{5.31}$$

Then it follows from the definition of $\hat{H}_x$ that

$$\hat{f}_x(x + td) - \hat{f}_x(x) < 0 \qquad \text{for all} \quad t \in (0, \varepsilon],$$

which means by Theorem 5.2.5(iv) that d is also a descent direction for f at x. On the other hand we also have

$$\hat{F}_x(x + td) = \max \left\{ \hat{F}_{i,x}(x + td) \mid i = 1, \dots, m_F \right\} < 0 \qquad \text{for all} \quad t \in (0, \varepsilon].$$

If $F_i(x) < 0$ for some $i = 1, \dots m_F$, then by the continuity of F_i there exists $\varepsilon_i > 0$ such that $F_i(x + td) \leq 0$ for all $t \in (0, \varepsilon_i]$ and if $F_i(x) = 0$ for some $i = 1, \dots, m_F$, then by (5.30) we have

$$\hat{F}_{i,x}(x + td) < \hat{F}_{i,x}(x) = F_i(x) = 0 \qquad \text{for all} \quad t \in (0, \varepsilon],$$

which means that d is a descent direction for $\hat{F}_{i,x}$ at x. Then by Theorem 5.2.5(iv) the direction d is also a descent for F_i and we have $\varepsilon_i' > 0$ such that

$$F_i(x + td) < F_i(x) = 0 \qquad \text{for all} \quad t \in (0, \varepsilon_i'].$$

Then by defining $\delta := \min_{i=1,\ldots,m_F} \{\varepsilon_i', \varepsilon, \delta'\}$ we get $x + td \in G$ for all $t \in (0, \delta]$ and so d is a feasible direction subject to G. □

To end this chapter we show how one can find a descent feasible direction for the linearized functions. Note that by Theorem 5.3.10 this direction is a feasible descent direction also for the original, improvement function.

Theorem 5.3.11. *Consider the problem*

$$\text{minimize} \quad \hat{H}_x(x + d) + \tfrac{1}{2}\|d\|^2 \quad \text{subject to} \quad x + d \in G_C. \tag{5.32}$$

The problem (5.32) has a unique solution $d^* = -\xi^*$*, where* $\xi^* = \arg\min \{\|\xi\| \mid \xi \in \partial \hat{H}_x(x)$ *and* $x - \xi \in G_C\}$.

Proof. This follows from the proof of Theorem 5.2.8. □

In this first part we have studied the first order generalized derivatives and first order optimality conditions. For second order analysis and optimality conditions we refer to **Auslender** (1982), **Ben-Tal and Zowe** (1982), **Burke** (1987), **Chaney** (1987a, 1987b, 1988), **Cominetti and Correa** (1990), **Hiriart-Urruty** (1983, 1984) **Hiriart-Urruty, Strodiot and Nguyen** (1984), **Hiriart-Urruty and Seeger** (1989), **Kawasaki** (1988), **Klatte and Tammer** (1988), **Pallascke, Recht and Urbański** (1987), **Rockafellar** (1985, 1989) and **Warga** (1984).

Part II

Nonsmooth Optimization

Chapter 1

Introduction

Consider the following nonlinear constrained optimization problem

$$(\mathcal{P}) \qquad \begin{cases} \text{minimize} & f(x) \\ \text{subject to} & x \in G, \end{cases}$$

where the objective function $f : \mathbf{R}^n \to \mathbf{R}$ is supposed to be a locally Lipschitz function on the feasible set $G \subset \mathbf{R}^n$. If f is continuously differentiable, then the problem $(\mathcal{P})$ is said to be *smooth* and if $G = \mathbf{R}^n$, then $(\mathcal{P})$ is said to be *unconstrained*. Further, if f is a convex function and G is a convex set, then $(\mathcal{P})$ is said to be *convex*.

Nonlinear optimization is one of the most important areas of applied mathematics and a great deal of research has been devoted to it during the last decades. Typically, optimization methods are iterative: starting from a given initial point $x_1 \in \mathbf{R}^n$, they construct a sequence $(x_i)_{i=1}^{\infty} \subset \mathbf{R}^n$ which is intended to converge to the required solution. A general iterative algorithm to solve the problem $(\mathcal{P})$ is as follows.

Algorithm 1.1 (Basic Algorithm).

Step 0: (Initialization). Find a feasible starting point $x_1 \in G$ and set $k := 1$.

Step 1: (Direction finding). Find a feasible descent direction $d_k \in \mathbf{R}^n$:

$$f(x_k + td_k) < f(x_k) \quad \text{and} \quad x_k + td_k \in G \quad \text{for some} \quad t > 0.$$

Step 2: (Stopping criterion). If x_k is "close enough" to the required solution then STOP.

Step 3: (Line search). Find a step size $t_k > 0$ such that

$$t_k \approx \arg \min_{t>0} \{ f(x_k + td_k) \} \quad \text{and} \quad x_k + t_k d_k \in G.$$

Step 4: (Updating). Set $x_{k+1} := x_k + t_k d_k$, $k := k + 1$ and go to Step 1.

93

The simplest optimization algorithms are constructed for smooth unconstrained versions of the problem $(\mathcal{P})$. The most efficient methods like conjugate gradient and (Quasi)-Newton methods employ derivative information from the problem functions. In Step 1 a descent direction may be generated by exploiting the fact that the direction opposite to the gradient is locally the steepest descent direction. Due to the necessary condition for a local optimum the gradient must be zero at each local solution, and by continuity it becomes small as soon as we are close to an optimal point. This fact may yield a stopping criterion in Step 2. In methods for smooth problems the line searches usually employ some efficient univariate smooth optimization method or some polynomial interpolation.

Nonsmoothness creates quite a lot of difficulties and requires additional work at almost every step. The first and hardest problem is in Step 1, since the direction opposite to an arbitrary subgradient need not be one of descent. This fact forces us also to modify the classical line search operations. Also, the stopping criterion is not clear anymore. The simplest counterexample in the nonsmooth case is the absolute-value function on the reals, which attains its global minimum at zero although, for example, plus one is a suitable nonzero subgradient at that minimum point and in its neighborhood.

The methods for nonsmooth optimization can be divided into two main classes: subgradient methods and bundle methods. They are based on the assumption that the problem functions are locally Lipschitz continuous and we can evaluate each function and its arbitrary subgradient at each point. These assumptions have proved to be quite natural in practice. Note that the problem functions need not be differentiable or convex.

The history of subgradient methods starts in the 60s. Their basic idea is to generalize the methods for smooth problems by replacing the gradient by an arbitrary subgradient. This simple idea poses two critical questions: how can we choose the step size and is there any implementable stopping rule? Several different proposals have been given to answer these questions, but we can state that the lack of an implementable stopping criterion is the main handicap of subgradient methods. Also the fact that the direction opposite to an arbitrary subgradient need not yield descent means that the subgradient methods are not descent methods; a descent method should produce at each iteration a new iteration point at which the function value is not greater than that at the previous point. The subgradient methods were mainly developed in the Soviet Union and an excellent overview can be found in **Shor** (1985).

At the moment the most promising methods in nonsmooth optimization seem to be the bundle methods. The guiding principle behind them is to exploit the previous iterations by gathering the subgradient information into a *bundle*. The pioneering bundle method, the ε-steepest descent method was developed in **Lemaréchal** (1976) and it was based on the conjugate subgradient method by **Lemaréchal** (1975) and **Wolfe** (1975). The well-known and widely used Fortran-codes M1FC1 and M2FC1 by Lemaréchal are employing this ε-steepest descent idea.

In his book (**Kiwiel** (1985)) Kiwiel gave a new approach to the bundle methods, which was based on the classical cutting plane method developed by **Kelley** (1960) and **Cheney and Goldstein** (1959). The basic idea in this generalized cutting plane method is to form a convex piecewise linear approximation to the objective function using the linearizations generated by subgradients. Kiwiel also presented two strategies to bound the number of stored subgradients: subgradient selection and aggregation. In spite of different backgrounds both methods (Lemaréchal's and Kiwiel's) generate the search direction at each iteration by solving quadratic direction finding problems, which are closely related.

The main difficulty in Lemaréchal's method is the a priori choice of an approximation tolerance which controls the radius of the ball in which the bundle model is thought to be a good approximation to the objective function. In turn, the greatest disadvantage of Kiwiel's method is its sensitivity to the scaling of the objective function. Also the uncertain line search operations may require, in general, many function evaluations compared with the number of iterations. The latest developments in nonsmooth optimization have been motivated by the need to overcome the difficulties described above. The Bundle Trust Region method developed by **Schramm** (1989), **Zowe** (1988), **Schramm and Zowe** (1990) and the Proximal Bundle method by **Kiwiel** (1990) combine the bundle idea with the classical Trust Region method by **Fletcher** (1987) and **Yuan** (1985). There exist strong similarities between these methods and from a practical point of view they proceed with approximately the same algorithm that differs only in technical details.

The most recent advance in the area of bundle methods has been presented in **Kiwiel** (1990), where the Tilted Proximal Bundle method was proposed. The guiding principle behind this method is the tilting of linearizations in order to cut off parts of the epigraph of the objective function. This idea tries to extract some second order information on the objective. Also some other recent modifications can be found in **Gaudioso and Monaco** (1991) and in **Mäkelä** (1990 a).

In what follows we give a short derivation and development of these bundle methods. We concentrate mainly on ways in which they generate descent directions. For simplicity we restrict ourselves only to the convex unconstrained version of problem ($\mathcal{P}$). After that we shall give a derivation of our method which generalizes the Proximal Bundle method to the constrained nonconvex case. It is based on the method derived in **Kiwiel** (1990) for convex unconstrained optimization. To show the reliability of our implementation some numerical experiments with the well-known test problems from literature are reported.

Chapter 2

A Survey of Bundle Methods

2.1. Preliminaries

Consider the unconstrained convex problem

$$(\mathcal{P}) \qquad \begin{cases} \text{minimize} & f(x) \\ \text{subject to} & x \in \mathbf{R}^n. \end{cases}$$

We are working under the following assumption: at each point $x \in \mathbf{R}^n$ we can evaluate

- at least one subgradient $\xi \in \partial f(x)$ and the function value $f(x)$.

The following two features are characteristic to bundle methods:

- the gathering of subgradient information from past iterations into a bundle;

- the concept of *serious* and *null steps* in line search: let $y_{k+1} := x_k + t_k d_k$ for some $t_k > 0$ and $\xi_{k+1} \in \partial f(y_{k+1})$. Then if

$$f(y_{k+1}) \leq f(x_k) - \delta_k \quad \text{for some} \quad \delta_k > 0,$$

    (i) make a *serious step*: $x_{k+1} := y_{k+1}$,

    (ii) otherwise make a *null step*: $x_{k+1} := x_k$.

In both cases add ξ_{k+1} into bundle.

We shall now give a short derivation of bundle methods for problem $(\mathcal{P})$. To see the relationships to classical methods we recall first some methods for smooth problems. Since the methods mainly differ in the strategies for generating the descent directions we shall concentrate on this question. To recall some difficulties arising in nonsmooth optimization we give also a short description of subgradient methods.

2.1.1. Smooth Minimization Methods

Suppose first that the objective function f is twice continuously differentiable and the point $x_k \in \mathbf{R}^n$ is nonoptimal, i.e. $\nabla f(x_k) \neq 0$ and there exists $d \in \mathbf{R}^n$ such that $f'(x_k; d) < 0$. Hence our aim is to find a direction d_k such that

$$f(x_k + d_k) < f(x_k). \tag{2.1}$$

We rewrite the problem $(\mathcal{P})$ in form

$$\begin{cases} \text{minimize} & f(x_k + d) - f(x_k) \\ \text{subject to} & d \in \mathbf{R}^n, \end{cases} \tag{2.2}$$

which has a solution $d_k \neq 0$. Due to Taylor's first order expansion

$$f(x_k + d) - f(x_k) = f'(x_k; d) + \|d\|\varepsilon(d), \tag{2.3}$$

where $\varepsilon(d) \to 0$ as $\|d\| \to 0$. By using a truncated form of (2.3) and the identity $f'(x_k; d) = \nabla f(x_k)^T d$ we obtain the following two equivalent approximate versions of (2.2)

$$\begin{cases} \text{minimize} & f'(x_k; d) \\ \text{subject to} & \|d\| \leq 1 \end{cases} \Longleftrightarrow \begin{cases} \text{minimize} & \nabla f(x_k)^T d \\ \text{subject to} & \|d\| \leq 1, \end{cases} \tag{2.4}$$

where the additional constraint $\|d\| \leq 1$ becomes necessary, since the function $f'(x_k; \cdot)$ is positively homogeneous. Clearly the latter minimization problem has the unique solution $d_k = -\nabla f(x_k)/\|\nabla f(x_k)\|$. This leads to the *steepest descent (gradient) method*

(Steepest Descent) $\boxed{d_k = -\nabla f(x_k).}$

Although the direction opposite of the gradient locally is the direction of steepest descent, suffers this simple method from low rate of convergence and so called siksak -phenomena. To improve the rate of convergence we employ the past direction to derive the *conjugate gradient method*

(Conjugate Gradient) $\boxed{d_k = -\nabla f(x_k) + \lambda_k d_{k-1},}$

where λ_k is defined for example by

$$\begin{cases} \lambda_0 = 0 \qquad \text{and} \\ \lambda_k = \dfrac{\|f(x_k)\|^2}{\|f(x_{k-1})\|^2} \quad \text{for} \quad k > 0. \end{cases}$$

The conjugate gradient method is a rather efficient method and suitable especially for large scale problems. To get some second order information we can employ a more accurate version of Taylor's expansion

$$f(x_k + d) - f(x_k) = \nabla f(x_k)^T d + \tfrac{1}{2} d^T \nabla^2 f(x_k) d + \|d\|^2 \varepsilon(d),$$

which leads to the problem

$$\begin{cases} \text{minimize} & \nabla f(x_k)^T d + \frac{1}{2} d^T \nabla^2 f(x_k) d \\ \text{subject to} & d \in \mathbf{R}^n. \end{cases} \qquad (2.5)$$

The necessary and sufficient condition for d to solve (2.5) is

$$\nabla_d \left(\nabla f(x_k)^T d + \frac{1}{2} d^T \nabla^2 f(x_k) d \right) = \nabla f(x_k) + \nabla^2 f(x_k) d = 0.$$

This establishes *Newton's method* (assuming that $\nabla^2 f(x_k)$ is nonsingular)

(Newton) $\boxed{d_k = -\nabla^2 f(x_k)^{-1} \nabla f(x_k).}$

Since the calculation of the Hessian inverse matrix $\nabla^2 f(x_k)^{-1}$ may be highly complicated we may try to approximate it by some $n \times n$ matrix H_k, i.e.

$$H_k \approx \nabla^2 f(x_k)^{-1}$$

and we obtain the *Quasi-Newton (variable metric) method*

(Quasi-Newton) $\boxed{d_k = -H_k \nabla f(x_k).}$

For updating the matrix H_k we can use the well-known DFP or BFGS -formulas (see for example **Fletcher** (1987)).

2.1.2. Subgradient Methods

To point out some difficulties arising with nonsmoothness we shall give short overview of subgradient methods, the pioneering methods in nonsmooth optimization. A more detailed presentation can be found in **Shor** (1985), **Lemaréchal** (1989b) and **Zowe** (1985).

Let us return back to the nonsmooth problem

$(\mathcal{P})$ $\quad \begin{cases} \text{minimize} & f(x) \\ \text{subject to} & x \in \mathbf{R}^n. \end{cases}$

Instead of the gradient $\nabla f(x_k)$ we can now employ only one subgradient $\xi_k \in \partial f(x_k)$. Hence the natural generalization of gradient method is to replace the gradient by the normalized subgradient to obtain

(Subgradient) $\boxed{d_k = -\xi_k / \|\xi_k\|.}$

As mentioned in the introduction the above strategy of generating d_k need not ensure descent and hence minimizing line searches become unrealistic. Also the standard stopping criterion can no longer be applied, since an arbitrary subgradient contains no information on the optimality condition $0 \in \partial f(x)$.

Due to these facts we are forced to use an a priori choice of step sizes t_k to avoid line searches and the stopping criterion. Thus we define the next iteration point by

$$\begin{cases} x_{k+1} := x_k - t_k \dfrac{\xi_k}{\|\xi_k\|} & \text{where} \quad \xi_k \in \partial f(x_k) \\ t_k > 0 \quad \text{(suitable).} \end{cases}$$

The following two lemmas justify some choices of step sizes.

Lemma 2.1.2.1. *Let x^* be a solution of $(\mathcal{P})$ and suppose that x_k is not a solution. Then*

$$\|x_{k+1} - x^*\| < \|x_k - x^*\|$$

whenever

$$0 < t_k < 2[f(x_k) - f(x^*)]/\|\xi_k\|.$$

Proof. See **Lemaréchal** (1989b) p. 544.   □

Lemma 2.1.2.2. *At each iteration k we have*

$$\|x_0 - x_k\| \le \sum_{j=0}^{k} t_k.$$

Proof. Elementary by the triangle inequality; see **Lemaréchal** (1989b) p. 545.   □

Due to these results we require to guarantee global convergence that

$$t_k \downarrow 0 \quad \text{when} \quad k \to \infty \quad \text{and} \quad \sum_{j=0}^{\infty} t_k = \infty.$$

In order to accelerate the rate of convergence we may try to generalize more efficient smooth methods than gradient method. One suitable method class are Quasi-Newton methods. Replacing the gradient by the normalized subgradient in Quasi-Newton method we obtain

(Space Dilation) $$\boxed{d_k = -H_k \xi_k / \xi_k^T H_k \xi_k^{1/2}.}$$

However the direct generalization by employing the standard DFP and BFGS up-dating formulas for the approximative Hessian inverse matrix H_k perform very poor numerical results see **Lemaréchal** (1982). These difficulties have led to the appearance of new ideas in the construction of generalized Quasi-Newton meth-ods. The most efficient methods at the moment are ellipsoid and space dilation methods by **Shor** (1985) and variable metric method by **Uryas'ev** (1991). As a conclusion about subgradient methods we may state that they suffer from quite poor theoretical convergence results, but may be applied quite efficiently in certain special cases.

2.2. Lemaréchal's $\mathcal{E}$-steepest Descent Method

As were seen in context of subgradient methods the direct generalization of steepest descent method did not give satisfactory results. The next natural at-tempt is to generalize the whole derivation of the gradient method by starting from the formulas (2.4). However the counterexample in **Wolfe** (1975) shows that this idea may cause the algorithm to take infinitely many steps without a significant decrease of the objective function and hence convergence to a nonopti-mal kink may occur. This disadvantage can be avoided by replacing the ordinary directional derivative by ε-directional derivative.

2.2.1. Theoretical Foundation

Instead of (2.1) our aim is now to find $d_k \in \mathbf{R}^n$ such that

$$f(x_k + d_k) < f(x_k) - \varepsilon \qquad \text{for some} \quad \varepsilon > 0. \tag{2.6}$$

Suppose that the point $x_k \in \mathbf{R}^n$ is not ε-optimal, in other words

$$0 \notin \partial_\varepsilon f(x_k). \tag{2.7}$$

Then the problem

$$\begin{cases} \text{minimize} & f'_\varepsilon(x_k; d) \\ \text{subject to} & \|d\| \leq 1 \end{cases} \tag{2.8}$$

has a solution. By using the identity

$$f'_\varepsilon(x_k; d) = \max\{\xi^\mathrm{T} d \mid \xi \in \partial_\varepsilon f(x_k)\}$$

we rewrite (2.8) in the form

$$\min_{\|d\| \le 1} \max_{\xi \in \partial_\varepsilon f(x_k)} \xi^T d.$$

Hence, by the well-known Minmax Theorem we can change the optimization order and we get

$$\max_{\xi \in \partial_\varepsilon f(x_k)} \min_{\|d\| \le 1} \xi^T d.$$

If $\xi \in \partial_\varepsilon f(x_k)$, then due to (2.7) $\xi \ne 0$ and the solution of the latter minimization problem is $d = -\xi/\|\xi\|$, so we have

$$\max_{\xi \in \partial_\varepsilon f(x_k)} \min_{\|d\| \le 1} \xi^T d = \max_{\xi \in \partial_\varepsilon f(x_k)} \xi^T(-\xi/\|\xi\|) = -\min_{\xi \in \partial_\varepsilon f(x_k)} \|\xi\|. \qquad (2.9)$$

Hence, for solving (2.8) we have to study the minimum-norm problem (which is uniquely solvable since $\partial_\varepsilon f(x_k)$ is a nonempty closed convex set)

$$\begin{cases} \text{minimize} & \|\xi\| \\ \text{subject to} & \xi \in \partial_\varepsilon f(x_k). \end{cases} \qquad (2.10)$$

2.2.2. Implementation

Notice that the problem (2.10) demands the knowledge of the whole ε-subdifferential $\partial_\varepsilon f(x_k)$, which in practice is too big a requirement. For this reason we have to approximate it somehow. First we define the *linearization error* for $\xi \in \partial f(y)$ by (see Figure 2.1)

$$\alpha(x; y) := f(x) - f(y) - \xi^T(x - y) \qquad \text{for all} \quad x \in \mathbf{R}^n.$$

Notice that due to the definition of subgradient for convex function we have

$$\alpha(x; y) \ge 0 \qquad \text{for all} \quad x, y \in \mathbf{R}^n.$$

We suppose that in addition to the current iteration point x_k we have some auxiliary points $y_j \in \mathbf{R}^n$ and subgradients $\xi_j \in \partial f(y_j)$ for $j \in J_k$, where the index set J_k is such that $\emptyset \ne J_k \subset \{1, \ldots, k\}$. We denote the linearization errors

$$\alpha_j^k := \alpha(x_k; y_j) \qquad \text{for all} \quad j \in J_k.$$

Based on Theorem I.3.2.16 we define for ε_k an approximation of $\partial_{\varepsilon_k} f(x_k)$ by

$$G_k(\varepsilon_k) = \{\xi \in \mathbf{R}^n \mid \xi = \sum_{j \in J_k} \lambda_j \xi_j, \ \sum_{j \in J_k} \lambda_j \alpha_j^k \le \varepsilon_k, \ \lambda_j \ge 0, \ \sum_{j \in J_k} \lambda_j = 1\}.$$

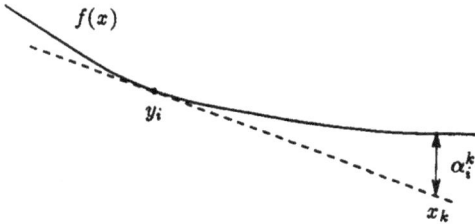

Figure 2.1. The linearization error

Lemma 2.2.2.1. *The set $G_k(\varepsilon_k)$ is convex and compact such that*

$$G_k(\varepsilon_k) \subset \partial_{\varepsilon_k} f(x_k).$$

Proof. By construction $G_k(\varepsilon_k)$ is convex and compact. Suppose that $\xi \in G_k(\varepsilon_k)$, then $\xi = \sum_{j \in J_k} \lambda_j \xi_j$ and $\xi_j \in \partial f(y_j)$. For all $x \in \mathbf{R}^n$ we have

$$
\begin{aligned}
f(x) = \sum_{j \in J_k} \lambda_j f(x) &\geq \sum_{j \in J_k} \lambda_j [f(y_j) + \xi_j^{\mathrm{T}}(x - y_j)] \\
&= \sum_{j \in J_k} \lambda_j [f(x_k) + \xi_j^{\mathrm{T}}(x - x_k) - (f(x_k) - f(y_j) - \xi_j^{\mathrm{T}}(x_k - y_j))] \\
&= \sum_{j \in J_k} \lambda_j [f(x_k) + \xi_j^{\mathrm{T}}(x - x_k) - \alpha_j^k] \\
&= \sum_{j \in J_k} \lambda_j f(x_k) + (\sum_{j \in J_k} \lambda_j \xi_j)^{\mathrm{T}}(x - x_k) - \sum_{j \in J_k} \lambda_j \alpha_j^k \\
&\geq f(x_k) + \xi^{\mathrm{T}}(x - x_k) - \varepsilon_k,
\end{aligned}
$$

which implies due to the definition of ε-subdifferential that $\xi \in \partial_{\varepsilon_k} f(x_k)$. $\square$

Now we replace $\partial_{\varepsilon_k} f(x_k)$ by the set $G_k(\varepsilon_k)$ in (2.10) to derive

$$
\begin{cases}
\text{minimize} & \frac{1}{2}\|\xi\|^2 \\
\text{subject to} & \xi \in G_k(\varepsilon_k).
\end{cases}
\tag{2.11}
$$

Due to the definition of $G_k(\varepsilon_k)$ this is equivalent to finding the multipliers λ_j for $j \in J_k$ that solve the quadratic problem

(B)
$$
\begin{cases}
\text{minimize} & \frac{1}{2}\|\sum_{j \in J_k} \lambda_j \xi_j\|^2 \\[2mm]
\text{subject to} & \sum_{j \in J_k} \lambda_j \alpha_j^k \le \varepsilon_k \\[2mm]
& \sum_{j \in J_k} \lambda_j = 1 \\[2mm]
\text{and} & \lambda_j \ge 0 \quad \text{for all} \quad j \in J_k.
\end{cases}
$$

If the multipliers λ_j^k solve problem (B), then due to (2.9) we obtain the search direction by

$$
d_k = -\sum_{j \in J_k} \lambda_j^k \xi_j. \tag{2.12}
$$

Remark 2.2.2.2. *The main difficulty in the ε-steepest descent method is the choice of the approximation tolerance ε_k in (B). This tolerance controls the radius of the ball in which the bundle model is thought to be a good approximation to the objective function. It typically depends highly on the situation and there exists a conflict between the sizes of ε_k, since*

(i) *for large ε_k the bundle $G_k(\varepsilon_k)$ gives a bad approximation of $\partial_{\varepsilon_k} f(x_k)$;*

(ii) *for small ε_k we will expect a small decrease of f due to (2.6).*

For these reasons it is difficult to derive exact rules for updating ε_k in the general case.

2.3. Kiwiel's Generalized Cutting Plane Method

2.3.1. The Cutting Plane Method

Let $\xi \in \partial f(y)$ and let us define the *linearization* $\bar{f}(x; \cdot) : \mathbf{R}^n \to \mathbf{R}$ such that

$$
\bar{f}(x; y) := f(y) + \xi^{\mathrm{T}}(x - y) \qquad \text{for all } x \in \mathbf{R}^n. \tag{2.13}
$$

Note that

$$
\alpha(x; y) = f(x) - \bar{f}(x; y) \ge 0. \tag{2.14}
$$

Due to the Theorem I.5.2.3(i) the convex function f has the representation

$$
f(x) = \max\{\bar{f}(x; y) \mid \xi \in \partial f(y), \ y \in \mathbf{R}^n\} \qquad \text{for all } x \in \mathbf{R}^n.
$$

However, for this representation we need the whole subdifferential $\partial f(y)$ and we again have to approximate it somehow. By using the same auxiliary points $y_j \in \mathbf{R}^n$ and subgradients $\xi_j \in \partial f(y_j)$ for $j \in J_k$ as before we define a piecewise linear approximation to f at the current iteration k by

$$\hat{f}^k(x) := \max \{\bar{f}(x; y_j) \mid j \in J_k\} \qquad \text{for all} \quad x \in \mathbf{R}^n. \tag{2.15}$$

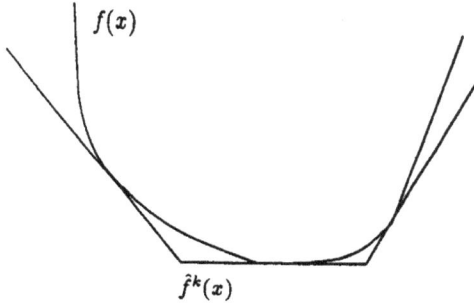

$f(x)$

$\hat{f}^k(x)$

Figure 2.2. The cutting plane model

Lemma 2.3.1.1. *For all $x \in \mathbf{R}^n$ and $j \in J_k$ we have*

(i) $\hat{f}^k(x) \leq f(x)$,

(ii) $\hat{f}^k(y_j) = f(y_j)$ *and*

(iii) $\hat{f}^k(x) = \max\limits_{j \in J_k} \{-\alpha(x_k; y_j) + \xi_j^T(x - x_k)\} + f(x_k)$.

Proof. The assertions (i) and (ii) follow directly from (2.14) and definition (2.15). To see (iii) notice that

$$\begin{aligned}
\bar{f}(x; y_j) &= f(y_j) + \xi_j^T(x - y_j) \\
&= -(f(x_k) - f(y_j) - \xi_j^T(x_k - y_j)) + \xi_j^T(x - x_k) + f(x_k) \\
&= -\alpha(x_k; y_j) + \xi_j^T(x - x_k) + f(x_k).
\end{aligned}$$

$\qquad\qquad\qquad\qquad\qquad\qquad\qquad\qquad\qquad\qquad\qquad\qquad\qquad\qquad\qquad\square$

The idea of the cutting plane method is to replace f by $\hat{f}^k$ to get the following approximation to problem (2.2)

$$\begin{cases} \text{minimize} & \hat{f}^k(x_k + d) - f(x_k) \\ \text{subject to} & d \in \mathbf{R}^n. \end{cases} \tag{2.16}$$

Note that problem (2.16) is still an unconstrained nonsmooth problem. Due to Lemma 2.3.1.1(iii) we can modify it as follows

$$\min_{d \in \mathbf{R}^n} \hat{f}^k(x_k + d) - f(x_k) = \min_{d \in \mathbf{R}^n} \max_{j \in J_k} \left\{ -\alpha_j^k + \xi_j^T d \right\}.$$

Finally, due to the minmax construction (2.16) can equivalently be transformed to finding a solution $(d, v) \in \mathbf{R}^{n+1}$ to a linearly constrained smooth *cutting plane* problem

(CP) $\begin{cases} \text{minimize} & v \\ \text{subject to} & -\alpha_j^k + \xi_j^T d \le v \quad \text{for all} \quad j \in J_k. \end{cases}$

2.3.2. A Generalization of the Cutting Plane Method

There are two main disadvantages in the original cutting plane method, namely the problem (CP) needs not have a solution and generally the method attains rather poor convergence results in practice. To avoid these drawbacks the following generalization was proposed in **Kiwiel** (1985). We add a regularizing penalty term $\frac{1}{2}\|d\|^2$ to the objective of (2.16) to guarantee a unique solution and thus we get the modified problem

$$\begin{cases} \text{minimize} & \hat{f}^k(x_k + d) + \frac{1}{2}\|d\|^2 - f(x_k) \\ \text{subject to} & d \in \mathbf{R}^n, \end{cases} \tag{2.17}$$

which can be converted as before to the smooth problem

(GCP) $\begin{cases} \text{minimize} & v + \frac{1}{2}\|d\|^2 \\ \text{subject to} & -\alpha_j^k + \xi_j^T d \le v \quad \text{for all} \quad j \in J_k. \end{cases}$

By duality this is equivalent to finding multipliers λ_j^k for $j \in J_k$ that solve the quadratic problem

(DCP) $\begin{cases} \text{minimize} & \frac{1}{2}\| \sum_{j \in J_k} \lambda_j \xi_j \|^2 + \sum_{j \in J_k} \lambda_j \alpha_j^k \\ \text{subject to} & \sum_{j \in J_k} \lambda_j = 1 \\ \text{and} & \lambda_j \ge 0 \quad \text{for all} \quad j \in J_k. \end{cases}$

If the multipliers λ_j^k solve the problem (DCP), then we obtain the search direction by

$$d_k = - \sum_{j \in J_k} \lambda_j^k \xi_j. \tag{2.18}$$

Remark 2.3.2.1. *In spite of totally different backgrounds and derivations the final direction finding problems (B) and (DCP) have a very close connection:*

- *If λ^k is an optimal solution of (DCP), then it is also an optimal solution of (B) for*

$$\varepsilon_k = \sum_{j \in J_k} \lambda_j^k \alpha_j^k.$$

Note that in (DCP) we do not need to choose the approximation tolerance ε_k, but this subproblem is rather sensitive to the scaling of the objective function (i.e. multiplication of f by a positive constant).

2.4. Proximal Bundle and Bundle Trust Region Methods

To avoid the drawbacks of the proceeding methods discussed in Remarks 2.2.2.2 and 2.3.3.1 the following two methods were developed independently: the Proximal Bundle (PB) method in **Kiwiel** (1990) and the Bundle Trust Region (BT) method in **Schramm** (1989) and **Zowe** (1988).

Instead of adding the regularizing penalty term $\frac{1}{2}\|d\|^2$ to the objective of (CP) we may use the idea of classical trust region methods to keep it as a constraint for the cutting plane model. Let $\sigma_k > 0$ and consider the following modification of (2.17)

$$\begin{cases} \text{minimize} & \hat{f}^k(x_k + d) - f(x_k) \\ \text{subject to} & \frac{1}{2}\|d\|^2 \leq \sigma_k, \end{cases} \tag{2.19}$$

which leads to

$$\begin{cases} \text{minimize} & v \\ \text{subject to} & -\alpha_j^k + \xi_j^T d \leq v \quad \text{for all} \quad j \in J_k \\ & \frac{1}{2}\|d\|^2 \leq \sigma_k. \end{cases} \tag{2.20}$$

For numerical reasons the quadratic constraint is not very pleasant and thus we modify the problem again. Let $u_k > 0$ and consider the problem

$$\begin{cases} \text{minimize} & v + \frac{u_k}{2}\|d\|^2 \\ \text{subject to} & -\alpha_j^k + \xi_j^T d \leq v \quad \text{for all} \quad j \in J_k. \end{cases} \tag{2.21}$$

The problems (2.20) and (2.21) have the following relationship.

Lemma 2.4.1.

(i) If (v_k, d_k) is an optimal solution of (2.20) and $u(\sigma_k)$ the corresponding Lagrange multiplier of the constraint

$$\tfrac{1}{2}\|d\|^2 \le \sigma_k,$$

then (v_k, d_k) is also an optimal solution of (2.21) for $u_k = u(\sigma_k)$.

(ii) If (v_k, d_k) is an optimal solution of (2.21), then it is also an optimal solution of (2.20) for

$$\sigma_k = \tfrac{1}{2}\|d_k\|^2.$$

Proof. See **Schramm** (1989) pp. 18–19. □

The dual formulation of (2.21) is to find multipliers λ_j for $j \in J_k$ that solve the quadratic problem

(PBT)
$$\begin{cases} \text{minimize} & \tfrac{1}{2}\|\sum_{j \in J_k} \lambda_j \xi_j\|^2 + u_k \sum_{j \in J_k} \lambda_j \alpha_j^k \\ \text{subject to} & \sum_{j \in J_k} \lambda_j = 1 \\ \text{and} & \lambda_j \ge 0 \quad \text{for all} \quad j \in J_k. \end{cases}$$

If the multipliers λ_j^k solve the problem (PBT), then we obtain the search direction by

$$d_k = -\tfrac{1}{u_k} \sum_{j \in J_k} \lambda_j^k \xi_j. \tag{2.22}$$

The connection between (B) and (PBT) is the following (cf. Lemma 2.4.1).

Lemma 2.4.2.

(i) If λ^k is an optimal solution of (B) and $u(\varepsilon_k)$ the corresponding Lagrange multiplier of the constraint

$$\sum_{j \in J_k} \lambda_j \alpha_j^k \le \varepsilon_k,$$

then λ^k is also an optimal solution of (PBT) for $u_k = u(\varepsilon_k)$.

(ii) *If λ^k is an optimal solution of (PBT), then it is also an optimal solution of (B) for*

$$\varepsilon_k = \sum_{j \in J_k} \lambda_j^k \alpha_j^k.$$

Proof. See **Schramm** (1989) p. 18. ☐

Remark 2.4.3.

(i) *Kiwiel's Generalized Cutting Plane method (GCP) is a special case of (PBT) with $u_k \equiv 1$.*

(ii) *The main differences between PB- and BT- methods consist in strategies for updating the weight u_k.*

(iii) *PB- and BT-methods do not need potentially unreliable line searches, since the step size control is included in the control of the weight u_k.*

2.5. The Tilted Proximal Bundle Method

The most recent step in the development of bundle methods was made in **Kiwiel** (1991). The main idea is to use so-called tilted cutting planes in order to get some second-order information and to employ some features of interior point methods (see **Meggido and Shub** (1989)).

Let $\varkappa \in (0, 1]$, $\theta \in [0, 1 - \varkappa]$ and $\xi \in \partial f(y)$. Instead of (2.13) we define the *tilted linearization*

$$\bar{f}(x; y; \theta) := f(y) + (1 - \theta)\xi^T(x - y) \qquad \text{for all} \quad x \in \mathbf{R}^n. \tag{2.23}$$

Note that

$$\bar{f}(x; y; 0) = \bar{f}(x; y).$$

Since $\bar{f}(x; y; \theta)$ need no longer be a lower approximation and the original linearization error may be negative, we define also a new linearization error by (cf. (2.14))

$$\alpha(x; y; \theta) := \max \{ f(x) - \bar{f}(x; y; \theta), \, \varkappa \cdot \alpha(x; y) \}.$$

Notice that

$$\alpha(x; y; \theta) \geq 0 \tag{2.24}$$

due to $\varkappa > 0$. The following result gives a motivation to employ tilted approximations.

Lemma 2.5.1. *If f is strictly convex and quadratic with the minimizer x^*, then*

$$\bar{f}(x^*; y; \theta) \leq \bar{f}(x^*; y; \tfrac{1}{2}) = f(x^*) \qquad \text{for all} \quad y \in \mathbf{R}^n \quad \text{and} \quad \theta \in [0, \tfrac{1}{2}].$$

Proof. See **Tarasov and Popova** (1984) p. 287. □

Now let $\xi_j \in \partial f(y_j)$ for $j \in J_k$ and let $\theta_j^k \in [0, 1 - \varkappa]$ for $j \in J_k$ be the corresponding tilting coefficients. Then for all $x \in \mathbf{R}^n$ we define the following tilted approximation to f (cf. Lemma 2.3.1.1(iii))

$$\hat{f}^k(x; \theta^k) := \max_{j \in J_k} \left\{ -\alpha(x_k; y_j; \theta_j^k) + (1 - \theta_j^k)\xi_j^{\mathrm{T}}(x - x_k) \right\} + f(x_k).$$

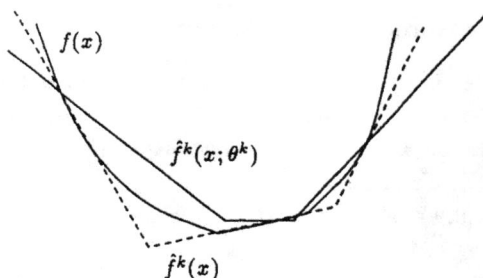

Figure 2.3. The tilted cutting plane model

Remark 2.5.2.

(i) *By choosing $\varkappa = 1$ we have*

$$\hat{f}^k(x; \theta^k) = \hat{f}^k(x; 0) = \hat{f}^k(x).$$

(ii) *By choosing $\theta = 0$ and dropping the term $\kappa \cdot \alpha(x; y)$ from the definition of $\alpha(x; y; \theta)$ we get the original tilted approximation proposed in* **Tarasov and Popova** *(1984).*

(iii) *A Tilted cutting plane is a real "cutting plane", since it cuts the parts of the epigraph of f while the standard cutting planes cuts nothing.*

Due to (2.23) $\hat{f}^k(x; \theta^k)$ is no longer a lower approximation of f, however (2.24) ensures that

$$\hat{f}^k(x_k; \theta^k) \leq f(x_k). \tag{2.25}$$

To exploit the proximal bundle idea we replace (2.21) with

$$\begin{cases} \text{minimize} \quad v + \frac{u_k}{2}\|d\|^2 \\ \text{subject to } -\alpha(x_k; y_j; \theta_j^k) + (1 - \theta_j^k)\xi_j^T d \leq v \quad \text{for all} \quad j \in J_k \end{cases} \tag{2.26}$$

and the dual formulation of (2.26) is to find multipliers λ_j^k for $j \in J_k$ that solve the problem

$$\text{(TPB)} \quad \begin{cases} \text{minimize} \quad \frac{1}{2}\|\sum_{j \in J_k} \lambda_j(1 - \theta_j^k)\xi_j\|^2 + u_k \sum_{j \in J_k} \lambda_j \alpha(x_k; y_j; \theta_j^k) \\ \text{subject to} \quad \sum_{j \in J_k} \lambda_j = 1 \\ \text{and} \quad \lambda_j \geq 0 \quad \text{for all} \quad j \in J_k. \end{cases}$$

If the multipliers λ_j^k solve the problem (TPB), then we obtain the search direction by

$$d_k = -\frac{1}{u_k} \sum_{j \in J_k} \lambda_j^k (1 - \theta_j^k)\xi_j. \tag{2.27}$$

For other survey type studies we refer to **Lemaréchal** (1989b), **Zowe** (1985) and **Yuan** (1987).

Chapter 3

Proximal Bundle Method for Nonconvex Constrained Optimization

Consider the following nonlinear constrained optimization problem

$$(\mathcal{P}) \quad \begin{cases} \text{minimize} & f(x) \\ \text{subject to} & F_i(x) \le 0 \quad \text{for} \quad i = 1, \ldots, m_F \\ & Cx \le b \\ & x_{min} \le x \le x_{max}. \end{cases}$$

where the objective function f and the constraint functions F_i for $i = 1, \ldots, m_F$ are real-valued locally Lipschitz functions defined on the Euclidean space $\mathbf{R}^n$, C is an $m_C \times n$ constraint matrix, b is an m_C-dimensional right-hand side vector and x_{min}, x_{max} are n-dimensional bound vectors. We denote by C_i for $i \in I = \{1, \ldots, m_C\}$ the rows of C. To simplify the notations we suppose that the simple bounds are included in the general linear constraints and thus we can omit them.

We suppose that the problem $(\mathcal{P})$ satisfies the Cottle constraint qualification and the feasible set $G := G_F \cap G_C$ is nonempty, where

$$G_F := \{x \in \mathbf{R}^n \mid F(x) \le 0\},$$
$$G_C := \{x \in \mathbf{R}^n \mid Cx \le b\}$$

and the total constraint function is defined by

$$F(x) = \max \{F_i(x) \mid i = 1, \ldots, m_F\} \quad \text{for all} \quad x \in \mathbf{R}^n. \tag{3.1}$$

Furthermore we suppose that at each $x \in \mathbf{R}^n$ we can evaluate

- subgradients $\xi^f \in \partial f(x)$, $\xi^F \in \partial F(x)$ and the function values $f(x)$, $F(x)$.

If the current iteration point $x_k \in \mathbf{R}^n$ is nonoptimal, then our aim is to find the descent direction $d_k \in \mathbf{R}^n$ which solves the problem

$$\begin{cases} \text{minimize} & f(x_k + d) - f(x_k) \\ \text{subject to} & x_k + d \in G. \end{cases} \tag{3.2}$$

By using the improvement function defined at $y \in \mathbf{R}^n$ by

$$H(x; y) = \max \{f(x) - f(y), \ F(x)\} \quad \text{for all} \quad x \in \mathbf{R}^n, \tag{3.3}$$

112

the generally constrained problem (3.2) can be modified as before to a linearly constrained problem

$$\begin{cases} \text{minimize} & H(x_k + d; x_k) \\ \text{subject to} & x_k + d \in G_C. \end{cases} \tag{3.4}$$

In what follows we shall construct a bundle method for solving the problem $(\mathcal{P})$. It is a generalization of the Proximal Bundle Method due to **Kiwiel** (1990) for the nonconvex constrained case. We shall go through the algorithm step by step and try to find answers to the questions raised by the nonsmoothness.

3.1. Direction Finding

We suppose for a while that the problem $(\mathcal{P})$ is *convex*. In subsection 3.1.3 we shall modify the direction finding step for nonconvex problem.

3.1.1. Derivation of the Direction Finding Problem

Suppose that the starting point x_1 is feasible and at the k-th iteration of the algorithm we have the current iteration point $x_k \in \mathbf{R}^n$, some auxiliary points $y_j \in \mathbf{R}^n$ and subgradients

$$\begin{aligned} \xi_j^f \in \partial f(y_j) \quad & \text{for} \quad j \in J_f^k \subset \{1, \ldots, k\} \quad \text{and} \\ \xi_j^F \in \partial F(y_j) \quad & \text{for} \quad j \in J_F^k \subset \{1, \ldots, k\}, \end{aligned}$$

where the index sets J_f^k and J_F^k are assumed to be nonempty. As before we define the linearizations at $x \in \mathbf{R}^n$ by

$$\begin{cases} \bar{f}_j(x) := \bar{f}(x; y_j) = f(y_j) + (\xi_j^f)^{\mathrm{T}}(x - y_j) & \text{for all} \quad j \in J_f^k \quad \text{and} \\ \bar{F}_j(x) := \bar{F}(x; y_j) = F(y_j) + (\xi_j^F)^{\mathrm{T}}(x - y_j) & \text{for all} \quad j \in J_F^k. \end{cases} \tag{3.5}$$

Notice that we do not need to store the auxiliary points y_j, since by denoting $f_j^k := \bar{f}_j(x_k)$ and $F_j^k := \bar{f}_j(x_k)$ the linearizations have the representation

$$\begin{cases} \bar{f}_j(x) = f_j^k + (\xi_j^f)^{\mathrm{T}}(x - x_k) & \text{for all} \quad j \in J_f^k \quad \text{and} \\ \bar{F}_j(x) = F_j^k + (\xi_j^F)^{\mathrm{T}}(x - x_k) & \text{for all} \quad j \in J_F^k \end{cases} \tag{3.6}$$

and we obtain the following recursive updating formula

$$\begin{cases} f_j^{k+1} = f_j^k + (\xi_j^f)^{\mathrm{T}}(x_{k+1} - x_k) & \text{for all} \quad j \in J_f^k \quad \text{and} \\ F_j^{k+1} = F_j^k + (\xi_j^F)^{\mathrm{T}}(x_{k+1} - x_k) & \text{for all} \quad j \in J_F^k. \end{cases} \tag{3.7}$$

Furthermore, for all $x \in \mathbb{R}^n$ we define again the polyhedral approximations by

$$\begin{cases} \hat{f}^k(x) := \max\{\bar{f}_j(x) \mid j \in J_f^k\} & \text{and} \\ \hat{F}^k(x) := \max\{\bar{F}_j(x) \mid j \in J_F^k\}. \end{cases} \tag{3.8}$$

Now we can define a convex piecewise linear approximation to the improvement function.

Definition 3.1.1.1. The *polyhedral approximation* to $H(\cdot\,;x_k)$ is defined by

$$\hat{H}^k(x) := \max\{\hat{f}^k(x) - f(x_k),\ \hat{F}^k(x)\} \quad \text{for all} \quad x \in \mathbb{R}^n.$$

Theorem 3.1.1.2. *The polyhedral approximation function $\hat{H}^k$ is convex. If in addition the problem is convex, then*

$$\hat{H}^k(x) \leq H(x;x_k) \quad \text{for all} \quad x \in \mathbb{R}^n.$$

Proof. As maxima of affine functions the polyhedral approximations $\hat{f}^k$ and $\hat{F}^k$ are convex, thus $\hat{H}^k$ is also convex as a maximum of two convex function. If the problem is convex then due to the definition of subgradient linearizations $\bar{f}_j$ and $\bar{F}_j$ are lower approximations of the problem functions. $\qquad\Box$

As in the classical cutting plane method we replace the original improvement function H by its polyhedral approximation function $\hat{H}^k$ and by employing the proximal bundle idea we obtain the following approximation to (3.4)

$$(\text{GCP}) \quad \begin{cases} \text{minimize} & \hat{H}^k(x_k + d) + \dfrac{u_k}{2}\|d\|^2 \\ \text{subject to} & C(x_k + d) \leq b. \end{cases}$$

To balance the presentation we shall use the following notations

$$\begin{cases} \alpha_{f,j}^k := f(x_k) - f_j^k & \text{for} \quad j \in J_f^k \\ \alpha_{F,j}^k := -F_j^k & \text{for} \quad j \in J_F^k \\ \alpha_{C,i}^k := -C_i^k := -C_i^{\mathrm{T}} x_k + b_i & \text{for} \quad i \in I. \end{cases} \tag{3.9}$$

By employing the same arguments as in chapter 2 we rewrite the problem (GCP) in the form

(BP)
$$
\begin{cases}
\text{minimize} & v + \dfrac{u_k}{2}\|d\|^2 \\[2mm]
\text{subject to} & -\alpha_{f,j}^k + (\xi_j^f)^{\mathrm{T}}d \le v \quad \text{for all} \quad j \in J_f^k \\[2mm]
& -\alpha_{F,j}^k + (\xi_j^F)^{\mathrm{T}}d \le v \quad \text{for all} \quad j \in J_F^k \\[2mm]
\text{and} & -\alpha_{C,i}^k + C_i^{\mathrm{T}}d \le 0 \quad \text{for all} \quad i \in I
\end{cases}
$$

and via dualization we find multipliers λ_j^k for $j \in J_f^k$, μ_j^k for $j \in J_F^k$ and ν_i^k for $i \in I$ that solve the problem

(DP)
$$
\begin{cases}
\text{minimize} & \dfrac{1}{2u_k}\Big\| \sum_{j \in J_f^k} \lambda_j \xi_j^f + \sum_{j \in J_F^k} \mu_j \xi_j^F + \sum_{i \in I} \nu_i C_i \Big\|^2 \\[3mm]
& + \sum_{j \in J_f^k} \lambda_j \alpha_{f,j}^k + \sum_{j \in J_F^k} \mu_j \alpha_{F,j}^k + \sum_{i \in I} \nu_i \alpha_{C,i}^k \\[3mm]
\text{subject to} & \sum_{j \in J_k} \lambda_j + \sum_{j \in J_F^k} \mu_j = 1 \\[3mm]
\text{and} & \lambda, \mu, \nu \ge 0.
\end{cases}
$$

Theorem 3.1.1.3.

(i) *Problems (GCP) and (BP) are equivalent.*

(ii) *Problem (BP) has a unique solution (d_k, v_k).*

(iii) *The vector $(d_k, v_k) \in \mathbf{R}^n \times \mathbf{R}$ solves problem (BP) if and only if there exist a vector $p_k \in \mathbf{R}^n$ and real Lagrange multipliers λ_j^k for $j \in J_f^k$, μ_j^k for $j \in J_F^k$ and ν_i^k for $i \in I$ such that*

(a) $\displaystyle \sum_{j \in J_f^k} \lambda_j^k + \sum_{j \in J_F^k} \mu_j^k = 1$ and $\lambda^k, \mu^k, \nu^k \ge 0$;

(b) $\lambda_j^k\,[-\alpha_{f,j}^k + (\xi_j^f)^{\mathrm{T}}d_k - v_k] = 0 \quad \text{for all} \quad j \in J_f^k$;

(c) $\mu_j^k\,[-\alpha_{F,j}^k + (\xi_j^F)^{\mathrm{T}}d_k - v_k] = 0 \quad \text{for all} \quad j \in J_F^k$;

(d) $\nu_i^k\,[-\alpha_{C,i}^k + C_i^{\mathrm{T}}d_k] = 0 \quad \text{for all} \quad i \in I$;

(e) $\displaystyle p_k = \sum_{j \in J_f^k} \lambda_j^k \xi_j^f + \sum_{j \in J_F^k} \mu_j^k \xi_j^F + \sum_{i \in I} \nu_i^k C_i$;

(f) $d_k = -\dfrac{1}{u_k} p_k$;

(g) $v_k = -\left[\dfrac{1}{u_k}\|p_k\|^2 + \displaystyle\sum_{j \in J_f^k} \lambda_j^k \alpha_{f,j}^k + \sum_{j \in J_F^k} \mu_j^k \alpha_{F,j}^k + \sum_{i \in I} \nu_i^k \alpha_{C,i}^k\right]$;

(h) $-\alpha_{f,j}^k + (\xi_j^f)^T d_k \le v_k \quad$ for all $\quad j \in J_f^k$;

(i) $-\alpha_{F,j}^k + (\xi_j^F)^T d_k \le v_k \quad$ for all $\quad j \in J_F^k$;

(j) $-\alpha_{C,i}^k + C_i^T d_k \le 0 \quad$ for all $\quad i \in I$.

(iv) *The multipliers λ^k, μ^k and ν^k satisfy the conditions (a)-(j) if and only if they solve problem (DP).*

Proof. It is evident that the problems (GCP) and (BP) are equivalent. Define a function $\rho : \mathbf{R}^n \to \mathbf{R}$ such that for all $d \in \mathbf{R}^n$

$$\rho(d) := \hat{H}^k(x_k + d) + \frac{u_k}{2}\|d\|^2.$$

Then problem (GCP) is equivalent to minimizing ρ over all $x_k + d \in G_C$. By Theorem 3.1.1.2 the function $\hat{H}^k$ is convex, which implies that ρ is strictly convex by the fact that the mapping $t \mapsto t^2$ is strictly convex. Since the constraint set G_C is nonempty, we derive that there exists a unique $d_k \in \mathbf{R}^n$ which minimizes ρ and by defining $v_k := \hat{H}^k(x_k + d_k)$ we deduce assertion (ii).

For assertion (iii) define the functions $q : \mathbf{R}^n \times \mathbf{R} \to \mathbf{R}$ by

$$q(d, v) := v + \frac{u_k}{2}\|d\|^2$$

and $Q_j : \mathbf{R}^n \times \mathbf{R} \to \mathbf{R}$ by

$$Q_j(d,v) := \begin{cases} -\alpha_{f,j}^k + (\xi_j^f)^T d - v & \text{for } j \in J_f^k, \\ -\alpha_{F,j}^k + (\xi_j^F)^T d - v & \text{for } j \in J_F^k, \\ -\alpha_{C,j}^k + C_j^T d & \text{for } j \in I. \end{cases}$$

Then by denoting $z = (d, v) \in \mathbf{R}^n \times \mathbf{R}$ the problem (BP) can equivalently be written in form

$$\text{minimize} \quad q(z) \quad \text{subject to} \quad Q_j(z) \le 0 \quad \text{for} \quad j \in J_f^k \cup J_F^k \cup I, \qquad (3.10)$$

where q is convex and Q_j are affine. For a point

$$(d, v) := (0, \max\{-\alpha_{f,j}^k, -\alpha_{F,j}^k \mid j \in J_f^k \cup J_F^k\})$$

we have $Q_j(d, u) \leq 0$ for all $j \in J_f^k \cup J_F^k \cup I$, i.e. (3.10) is consistent. Then by the optimality condition established in Corollary I.5.3.5 the fact that (d_k, v_k) solves the problem (3.10) is equivalent to the condition that there exist $\lambda_j^k \geq 0$ such that $\lambda_j^k Q_j(d_k, v_k) = 0$ for $j \in J_f^k$, $\mu_j^k \geq 0$ such that $\mu_j^k Q_j(d_k, v_k) = 0$ for $j \in J_F^k$, $\nu_i^k \geq 0$ such that $\nu_i^k Q_i(d_k, v_k) = 0$ for $i \in I$ and

$$0 \in \partial q(d_k, v_k) + \sum_{j \in J_f^k} \lambda_j^k \partial Q_j(d_k, v_k) + \sum_{j \in J_F^k} \mu_j^k \partial Q_j(d_k, v_k) + \sum_{i \in I} \nu_i^k \partial Q_i(d_k, v_k).$$

It is evident that the functions q and Q_j are continuously differentiable as affine functions and their gradients are $\nabla q(d, v) = (u_k d, 1)$, $\nabla Q_j(d, v) = (\xi_j^f, -1)$ for $j \in J_f^k$, $\nabla Q_j(d, v) = (\xi_j^F, -1)$ for $j \in J_F^k$ and $\nabla Q_i(d, v) = (C_i, 0)$ for $i \in I$. Thus we have

$$0 = (u_k d_k, 1) + \sum_{j \in J_f^k} \lambda_j^k(\xi_j^f, -1) + \sum_{j \in J_F^k} \mu_j^k(\xi_j^F, -1) + \sum_{i \in I} \nu_i^k(C_i, 0),$$

which implies parts (a)-(f). The parts (h)-(j) are evident, because the solution point (d_k, v_k) must also be feasible. For part (g) we sum up equations (b)-(d) for all $j \in J_f^k$, $j \in J_F^k$ and $i \in I$ to obtain

$$\Big(\sum_{j \in J_f^k} \lambda_j^k + \sum_{j \in J_F^k} \mu_j^k\Big) v_k = p_k^T d_k + \sum_{j \in J_f^k} \lambda_j^k \alpha_{f,j}^k + \sum_{j \in J_F^k} \mu_j^k \alpha_{F,j}^k + \sum_{i \in I} \nu_i^k \alpha_{C,i}^k,$$

which establishes assertion (ii). □

3.1.2. Subgradient Aggregation

The direction finding problem (BP) seems now to be suitable for generating a feasible descent direction. However, there is an open question, which we have not answered yet: how to choose the index sets J_f^k and J_F^k? As mentioned at the beginning of this chapter the index sets must be nonempty subsets of $\{1, \dots, k\}$. To guarantee progress of the algorithm it seems natural to demand that the current iteration index k belongs to the index sets. Thus the simplest way might be to choose directly

$$J_f^k := J_F^k := \{1, \dots, k\}. \tag{3.11}$$

However in practice this strategy presents serious problems with storage and computation after a large number of iterations. Note that at each iteration k the number of linear constraints in direction finding problem (BP) would be $2k + m_C$. For this reason we have to drop some extra constraints.

We shall now present the *subgradient aggregation* strategy (cf. **Kiwiel** (1985)) for dropping some constraints of the problem (BP). The idea is to aggregate the constraints generated by the past subgradients and thus to keep the number of constraints bounded.

Definition 3.1.2.1. Let λ_j^k for $j \in J_f^k$ and μ_j^k for $j \in J_F^k$ be the Lagrange multipliers of the problem (BP) at iteration k and denote $\lambda_f^k := \sum_{j \in J_f^k} \lambda_j^k$ and $\mu_F^k := \sum_{j \in J_k} \mu_j^k$. We define the *scaled multipliers* for all $j \in J_f^k$ and $j \in J_F^k$ by

$$\tilde{\lambda}_j^k := \begin{cases} \lambda_j^k/\lambda_f^k, & \text{if} \quad \lambda_f^k > 0 \\ 1/|J_f^k|, & \text{if} \quad \lambda_f^k = 0 \end{cases} \quad \text{and} \quad \tilde{\mu}_j^k := \begin{cases} \mu_j^k/\mu_F^k, & \text{if} \quad \mu_F^k > 0 \\ 1/|J_F^k|, & \text{if} \quad \mu_F^k = 0, \end{cases}$$

the *aggregate subgradients* by

$$(p_f^k, \tilde{f}_p^k, \tilde{\alpha}_{f,p}^k) := \sum_{j \in J_f^k} \tilde{\lambda}_j^k (\xi_j^f, f_j^k, \alpha_{f,j}^k) \text{ and } (p_F^k, \tilde{F}_p^k, \tilde{\alpha}_{F,p}^k) := \sum_{j \in J_F^k} \tilde{\mu}_j^k (\xi_j^F, F_j^k, \alpha_{F,j}^k)$$

and the *aggregate linearizations* by

$$\tilde{f}_p(x) := \tilde{f}_p^k + (p_f^k)^{\mathrm{T}}(x - x_k) \quad \text{and} \quad \tilde{F}_p(x) := \tilde{F}_p^k + (p_F^k)^{\mathrm{T}}(x - x_k).$$

Theorem 3.1.2.2. Let λ_j^k for $j \in J_f^k$, μ_j^k for $j \in J_F^k$ and ν_i^k for $i \in I$ be the Lagrange multipliers of the problem (BP) and denote $f_p^{k+1} := \tilde{f}_p(x_{k+1})$ and $F_p^{k+1} := \tilde{F}_p(x_{k+1})$. Then at iteration k we have

(i) $\displaystyle\sum_{j \in J_f^k} \tilde{\lambda}_j^k = 1, \ \sum_{j \in J_F^k} \tilde{\mu}_j^k = 1$ and $\lambda_f^k + \mu_F^k = 1$,

(ii) $\tilde{f}_p^k = \tilde{f}_p(x_k)$ and $\tilde{F}_p^k = \tilde{F}_p(x_k)$,

(iii) $f_p^{k+1} = \tilde{f}_p^k + (p_f^k)^{\mathrm{T}}(x_{k+1} - x_k)$ and $F_p^{k+1} = \tilde{F}_p^k + (p_F^k)^{\mathrm{T}}(x_{k+1} - x_k)$,

(iv) $\displaystyle\tilde{f}_p(x) = \sum_{j \in J_f^k} \tilde{\lambda}_j^k \tilde{f}_j(x)$ and $\tilde{F}_p(x) = \sum_{j \in J_F^k} \tilde{\mu}_j^k \tilde{F}_j(x)$,

(v) $\displaystyle p_k = \lambda_f^k p_f^k + \mu_F^k p_F^k + \sum_{i \in I} \nu_i^k C_i$,

(vi) $v_k = -\left[\dfrac{1}{u_k}\|p_k\|^2 + \lambda_f^k\tilde{\alpha}_{f,p}^k + \mu_F^k\tilde{\alpha}_{F,p}^k + \displaystyle\sum_{i\in I}\nu_i^k\alpha_{C,i}^k\right].$

Proof. The assertions (i)–(iii) follow directly from the above definition and for assertion (iv) we have

$$\tilde{f}_p(x) = \tilde{f}_p^k + (p_f^k)^T(x - x_k) = \sum_{j\in J_f^k}\tilde{\lambda}_j^k(f_j^k + (\xi_j^f)^T(x - x_k)) = \sum_{j\in J_f^k}\tilde{\lambda}_j^k\bar{f}_j(x)$$

$$\tilde{F}_p(x) = \tilde{F}_p^k + (p_F^k)^T(x - x_k) = \sum_{j\in J_F^k}\tilde{\lambda}_j^k(F_j^k + (\xi_j^F)^T(x - x_k)) = \sum_{j\in J_F^k}\tilde{\lambda}_j^k\bar{F}_j(x).$$

By definition and Theorem 3.1.1.3(iii) we have

$$\lambda_f^k p_f^k + \mu_F^k p_F^k + \sum_{i\in I}\nu_i^k C_i = \lambda_f^k\sum_{j\in J_f^k}\tilde{\lambda}_j^k\xi_j^f + \mu_F^k\sum_{j\in J_F^k}\tilde{\mu}_j^k\xi_j^F + \sum_{i\in I}\nu_i^k C_i$$

$$= \sum_{j\in J_f^k}\lambda_j^k\xi_j^f + \sum_{j\in J_F^k}\mu_j^k\xi_j^F + \sum_{i\in I}\nu_i^k C_i$$

$$= p_k,$$

which establishes assertion (v) and the proof of assertion (vi) is analogous. □

Lemma 3.1.2.3. *The problem (BP) is equivalent to the reduced problem*

(RP)
$$\begin{cases}
\text{minimize} & v + \dfrac{u_k}{2}\|d\|^2 \\[1mm]
\text{subject to} & -\alpha_{f,j}^k + (\xi_j^f)^T d \le v \quad \text{for all}\quad j\in\hat{J}_f^k \\[1mm]
& -\tilde{\alpha}_{f,p}^k + (p_f^k)^T d \le v \\[1mm]
& -\alpha_{F,j}^k + (\xi_j^F)^T d \le v \quad \text{for all}\quad j\in\hat{J}_F^k \\[1mm]
& -\tilde{\alpha}_{F,p}^k + (p_F^k)^T d \le v \\[1mm]
\text{and} & -\alpha_{C,i}^k + C_i^T d \le 0 \quad \text{for all}\quad i\in I
\end{cases}$$

where $\hat{J}_f^k$ and $\hat{J}_F^k$ are any subsets of J_f^k and J_F^k, respectively.

Proof. Define multipliers $\hat{\lambda}_p^k := \lambda_f^k$, $\hat{\mu}_p^k := \mu_F^k$, $\hat{\lambda}_j^k := 0$ for all $j\in\hat{J}_f^k$, $\hat{\mu}_j^k := 0$ for all $j\in\hat{J}_F^k$ and $\hat{\nu}_i^k := \nu_i^k$ for all $i\in I$. Due to Theorem 3.1.2.2 these multipliers satisfy the conditions (a)-(j) in Theorem 3.1.1.3 (iii), which implies the assertion. □

The problem (RP) seems to be suitable for keeping the index sets finite, since they can be chosen without any restrictions. However there is one drawback: at the beginning of iteration k the vectors p_f^k and p_F^k are unknown and we cannot use them in the aggregated constraints. This drawback can be eliminated by generating p_f^k and p_F^k recursively as follows. At the first iteration let $x_1 \in G$ be a feasible starting point supplied by the user, then we initialize our algorithm by

$$
\begin{cases}
y_1 := x_1, & p_f^0 := \xi_1^f \in \partial f(y_1), & f_p := f_1^1 := f(y_1) \\
J_1 := \{1\}, & p_F^0 := \xi_1^F \in \partial F(y_1), & F_p^1 := F_1^1 := F(y_1).
\end{cases}
\tag{3.12}
$$

At iteration k we replace the unknown vectors $\tilde{f}_p^k$, $\tilde{F}_p^k$, p_f^k and p_F^k by the previously generated f_p^k, F_p^k, p_f^{k-1} and p_F^{k-1}, respectively, and define

$$
\begin{cases}
\alpha_{f,p}^k := f(x_k) - f_p^k & \text{and} \\
\alpha_{F,p}^k := -F_p^k.
\end{cases}
\tag{3.13}
$$

This leads to the problem

$$
\text{(ABP)}
\begin{cases}
\text{minimize} & v + \dfrac{u_k}{2}\|d\|^2 \\
\text{subject to} & -\alpha_{f,j}^k + (\xi_j^f)^T d \le v \quad \text{for all} \quad j \in J_f^k \\
& -\alpha_{f,p}^k + (p_f^{k-1})^T d \le v \\
& -\alpha_{F,j}^k + (\xi_j^F)^T d \le v \quad \text{for all} \quad j \in J_F^k \\
& -\alpha_{F,p}^k + (p_F^{k-1})^T d \le v \\
\text{and} & -\alpha_{C,i}^k + C_i^T d \le 0 \quad \text{for all} \quad i \in I
\end{cases}
$$

and via dualization we find multipliers λ_p^k, μ_p^k, λ_j^k for $j \in J_f^k$, μ_j^k for $j \in J_F^k$ and ν_i^k for $i \in I$ that solve the problem (ADP)

$$
\begin{cases}
\text{minimize} & \dfrac{1}{2u_k}\left\| \sum_{j \in J_f^k} \lambda_j \xi_j^f + \lambda_p p_f^{k-1} + \sum_{j \in J_F^k} \mu_j \xi_j^F + \mu_p p_F^{k-1} + \sum_{i \in I} \nu_i C_i \right\|^2 \\
& + \sum_{j \in J_f^k} \lambda_j \alpha_{f,j}^k + \lambda_p \alpha_{f,p}^k + \sum_{j \in J_F^k} \mu_j \alpha_{F,j}^k + \mu_p \alpha_{F,p}^k + \sum_{i \in I} \nu_i \alpha_{C,i}^k \\
\text{subject to} & \sum_{j \in J_k} \lambda_j + \lambda_p + \sum_{j \in J_F^k} \mu_j + \mu_p = 1 \\
\text{and} & \lambda_j, \lambda_p, \mu_j, \mu_p, \nu_i \ge 0.
\end{cases}
$$

Corollary 3.1.2.4.

(i) *Problem (ABP) has a unique solution (d_k, v_k).*

(ii) *The vector $(d_k, v_k) \in \mathbf{R}^n \times \mathbf{R}$ solves problem (ABP) if and only if there exist a vector $p_k \in \mathbf{R}^n$ and real Lagrange multipliers λ_p^k, μ_p^k, λ_j^k for $j \in J_f^k$, μ_j^k for $j \in J_F^k$ and ν_i^k for $i \in I$ such that*

(a) $\displaystyle\sum_{j \in J_f^k} \lambda_j^k + \lambda_p^k + \sum_{j \in J_F^k} \mu_j^k + \mu_p^k = 1$ and $\lambda_p^k, \lambda_j^k, \mu_p^k, \mu_j^k, \nu_i^k \geq 0$;

(b) $\lambda_j^k [-\alpha_{f,j}^k + (\xi_j^f)^{\mathrm{T}} d_k - v_k] = 0 \quad$ for all $\quad j \in J_f^k$;

(c) $\lambda_p^k [-\alpha_{f,p}^k + (p_f^{k-1})^{\mathrm{T}} d_k - v_k] = 0$;

(d) $\mu_j^k [-\alpha_{F,j}^k + (\xi_j^F)^{\mathrm{T}} d_k - v_k] = 0 \quad$ for all $\quad j \in J_F^k$;

(e) $\mu_p^k [-\alpha_{F,p}^k + (p_F^{k-1})^{\mathrm{T}} d_k - v_k] = 0$;

(f) $\nu_i^k [-\alpha_{C,i}^k + C_i^{\mathrm{T}} d_k] = 0 \quad$ for all $\quad i \in I$;

(g) $\displaystyle p_k = \sum_{j \in J_f^k} \lambda_j^k \xi_j^f + \lambda_p^k p_f^{k-1} + \sum_{j \in J_F^k} \mu_j^k \xi_j^F + \mu_p^k p_F^{k-1} + \sum_{i \in I} \nu_i^k C_i$;

(h) $d_k = -\dfrac{1}{u_k} p_k$;

(i) $v_k = -\left[\dfrac{1}{u_k} \|p_k\|^2 + \displaystyle\sum_{j \in J_f^k} \lambda_j^k \alpha_{f,j}^k + \lambda_p^k \alpha_{f,p}^k + \sum_{j \in J_F^k} \mu_j^k \alpha_{F,j}^k + \mu_p^k \alpha_{F,p}^k + \sum_{i \in I} \nu_i^k \alpha_{C,i}^k\right]$;

(j) $-\alpha_{f,j}^k + (\xi_j^f)^{\mathrm{T}} d_k \leq v_k \quad$ for all $\quad j \in J_f^k$;

(k) $-\alpha_{f,p}^k + (p_f^{k-1})^{\mathrm{T}} d_k \leq v_k$;

(l) $-\alpha_{F,j}^k + (\xi_j^F)^{\mathrm{T}} d_k \leq v_k \quad$ for all $\quad j \in J_F^k$;

(m) $-\alpha_{F,p}^k + (p_F^{k-1})^{\mathrm{T}} d_k \leq v_k$;

(n) $-\alpha_{C,i}^k + C_i^{\mathrm{T}} d_k \leq 0 \quad$ for all $\quad i \in I$.

(iii) *The multipliers λ_j^k, λ_p^k, μ_j^k, μ_p^k and ν_i^k satisfy the conditions (a)-(n) if and only if they solve problem (ADP).*

Proof. Follows directly from Theorem 3.1.1.3. $\qquad\qquad\qquad\qquad\quad\square$

Suppose now that λ_p^k, μ_p^k, λ_j^k for $j \in J_f^k$, μ_j^k for $j \in J_F^k$ and ν_i^k for $i \in I$ are the Lagrange multipliers of the problem (ABP). Then we can similarly denote

$\lambda_f^k := \lambda_p^k + \sum_{j \in J_f^k} \lambda_j^k$ and $\mu_F^k := \mu_p^k + \sum_{j \in J_F^k} \mu_j^k$ in Definition 3.1.2.1 and define the scaled multipliers for all $j \in J_f^k$ and $j \in J_f^k$ by

$$\tilde{\lambda}_j^k := \begin{cases} \lambda_j^k/\lambda_f^k, & \text{if } \lambda_f^k > 0 \\ 1/(|J_f^k|+1), & \text{if } \lambda_f^k = 0 \end{cases}$$

$$\tilde{\lambda}_p^k := \begin{cases} \lambda_p^k/\lambda_f^k, & \text{if } \lambda_f^k > 0 \\ 1/(|J_f^k|+1), & \text{if } \lambda_f^k = 0 \end{cases}$$

$$\tilde{\mu}_j^k := \begin{cases} \mu_j^k/\mu_F^k, & \text{if } \mu_F^k > 0 \\ 1/(|J_F^k|+1), & \text{if } \mu_F^k = 0 \end{cases} \tag{3.14}$$

$$\tilde{\mu}_p^k := \begin{cases} \mu_p^k/\mu_F^k, & \text{if } \mu_F^k > 0 \\ 1/(|J_F^k|+1), & \text{if } \mu_F^k = 0, \end{cases}$$

and the aggregate subgradients by

$$(p_f^k, \tilde{f}_p^k) := \sum_{j \in J_f^k} \tilde{\lambda}_j^k(\xi_j^f, f_j^k) + \tilde{\lambda}_p^k(p_f^{k-1}, f_p^k) \quad \text{and}$$

$$(p_F^k, \tilde{F}_p^k) := \sum_{j \in J_F^k} \tilde{\mu}_j^k(\xi_j^F, F_j^k) + \tilde{\mu}_p^k(p_F^{k-1}, F_p^k). \tag{3.15}$$

Now we are allowed to choose the index sets J_f^k and J_f^k in problem (ABP) quite freely. In practice this choice has a strong effect on the trade-off between efficiency and amount of work per iteration. If the index sets are large, then each iteration is rather efficient but needs much storage and computing. In the opposite situation there is less work per iteration, but some iterations may not be very efficient and so we have to take very many iterations to achieve the required accuracy. To strike a balance we use a user-supplied bound $M_\xi \geq 2$ on the number of indices.

3.1.3. Nonconvexity

In this section we try to analyze the difficulties created by nonconvexity. One problem is that a nonconvex function may have several local minima. Thus in the nonconvex case we content ourselves with finding only one local minimum.

To see what other good properties convexity has, let us for a moment assume that all the problem functions are convex. By denoting

$$\tilde{\alpha}_p^k := \lambda_f^k \tilde{\alpha}_{f,p}^k + \mu_F^k \tilde{\alpha}_{F,p}^k$$

we obtain the following result

Theorem 3.1.3.1. *Suppose that all the problem functions are convex and $x_k \in G_F$. Then at each iteration k we have*

(i) $\xi_j^f \in \partial_{\alpha_{f,j}^k} f(x_k) \subset \partial_{\alpha_{f,j}^k} H(x_k; x_k)$ *for all $j \in J_f^k$;*

(ii) $p_f^{k-1} \in \partial_{\alpha_{f,p}^k} f(x_k) \subset \partial_{\alpha_{f,p}^k} H(x_k; x_k)$;

(iii) $p_f^k \in \partial_{\tilde{\alpha}_{f,p}^k} f(x_k) \subset \partial_{\tilde{\alpha}_{f,p}^k} H(x_k; x_k)$;

(iv) $\xi_j^F \in \partial_{\alpha_{F,j}^k} F(x_k) \subset \partial_{\alpha_{F,j}^k} H(x_k; x_k)$ *for all $j \in J_F^k$;*

(v) $p_F^{k-1} \in \partial_{\alpha_{F,p}^k} F(x_k) \subset \partial_{\alpha_{F,p}^k} H(x_k; x_k)$;

(vi) $p_F^k \in \partial_{\tilde{\alpha}_{F,p}^k} F(x_k) \subset \partial_{\tilde{\alpha}_{F,p}^k} H(x_k; x_k)$;

(vii) $\lambda_f^k p_f^k + \mu_F^k p_F^k \in \partial_{\tilde{\alpha}_p^k} H(x_k; x_k)$;

(viii) $p_k \in \partial_{\tilde{\alpha}_p^k} H(x_k; x_k) + \sum_{i \in I} \nu_i^k C_i$;

(ix) $\alpha_{f,j}^k, \alpha_{F,j}^k, \alpha_{f,p}^k, \alpha_{F,p}^k, \tilde{\alpha}_{f,p}^k, \tilde{\alpha}_{F,p}^k, \alpha_{C,i}^k, \tilde{\alpha}_p^k \geq 0$ *for all $j \in J_f^k$, $j \in J_F^k$ and $i \in I$.*

Proof. In assertions (i)–(vi) the latter inclusions follow directly from Theorem I.5.3.2. Further

$$f(x) \geq f_j^k + (\xi_j^f)^{\mathrm{T}}(x - x_k) = f(x_k) + (\xi_j^f)^{\mathrm{T}}(x - x_k) - \left[f(x_k) - f_j^k\right]$$
$$= f(x_k) + (\xi_j^f)^{\mathrm{T}}(x - x_k) - \alpha_{f,j}^k,$$

which implies by the definition of ε-subdifferential that $\xi_j^f \in \partial_{\alpha_{f,j}^k} f(x_k)$. By definition the aggregated subgradients p_f^k and p_f^{k-1} are convex combinations of the past subgradients ξ_j^f and f_p^k and $\tilde{f}_p^k$ are the corresponding convex combinations of f_j^k. Then at each iteration k there exist multipliers $\hat{\lambda}_j^k \geq 0$ such that $\sum_{j=1}^k \hat{\lambda}_j^k = 1$ and

$$(p_f^k, \tilde{f}_p^k) = \sum_{j=1}^k \hat{\lambda}_j^k (\xi_j^f, f_j^k) \tag{3.16}$$

and $\hat{\lambda}_j^{k-1} \geq 0$ such that $\sum_{j=1}^{k-1} \hat{\lambda}_j^{k-1} = 1$ and

$$(p_f^{k-1}, f_p^k) = \sum_{j=1}^{k-1} \hat{\lambda}_j^{k-1} (\xi_j^f, f_j^k). \tag{3.17}$$

Then by (3.16) and Theorem 3.1.1.2 for all $x \in \mathbf{R}^n$ we have

$$
\begin{aligned}
f(x) = \sum_{j=1}^{k} \hat{\lambda}_j^k f(x) &\geq \sum_{j=1}^{k} \hat{\lambda}_j^k f_j^k + \left(\sum_{j=1}^{k} \hat{\lambda}_j^k \xi_j^f \right)^{\mathrm{T}} (x - x_k) \\
&= \tilde{f}_p^k + (p_f^k)^{\mathrm{T}}(x - x_k) = f(x_k) + (p_f^k)^{\mathrm{T}}(x - x_k) - \left[f(x_k) - \tilde{f}_p^k \right] \\
&= f(x_k) + (p_f^k)^{\mathrm{T}}(x - x_k) - \tilde{\alpha}_{f,p}^k,
\end{aligned}
$$

which implies by the definition of subdifferential that $p_f^k \in \partial_{\tilde{\alpha}_{f,p}^k} f(x_k)$. The assertion (ii) follows then similarly from (3.17) and the proof of assertions (iv)-(vi) is similar if one uses the assumption that $F(x_k) \leq 0$. For (vii) we have by the proof of assertions (iii) and (vi) that

$$
\begin{aligned}
f(x) &\geq f(x_k) + (p_f^k)^{\mathrm{T}}(x - x_k) - \tilde{\alpha}_{f,p}^k \qquad \text{and} \\
F(x) &\geq (p_F^k)^{\mathrm{T}}(x - x_k) - \tilde{\alpha}_{F,p}^k.
\end{aligned}
$$

Then by the fact that $H(x_k; x_k) = 0$ we have

$$
\begin{aligned}
H(x; x_k) = \max \left\{ f(x) - f(x_k), F(x) \right\} &\geq \lambda_f^k [f(x) - f(x_k)] + \mu_F^k F(x) \\
&\geq (\lambda_f^k p_f^k + \mu_F^k p_F^k)^{\mathrm{T}}(x - x_k) - (\lambda_f^k \tilde{\alpha}_{f,p}^k + \mu_F^k \tilde{\alpha}_{F,p}^k) \\
&= H(x_k; x_k) + (\lambda_f^k p_f^k + \mu_F^k p_F^k)^{\mathrm{T}}(x - x_k) - \tilde{\alpha}_p^k,
\end{aligned}
$$

which implies assertion (vii) and assertion (viii) follows then directly from (vii) and Theorem 3.1.2.2(v). By setting $x = x_k$ in the preceding equations we get the assertion (viii).   □

By Theorem 3.1.1.2 the polyhedral approximations were lower approximations to the problem functions. Further, Theorem 3.1.3.1 means that the linearization errors are measures that indicate how much the subgradients calculated at trial points y_j deviate from being members of subdifferentials at x_k. It is easy to see that for strictly convex function the linearization error increases as we move away from a given trial point. This means that in the direction finding problem (ABP) we automatically weight the past local subgradients less heavily than the obsolete ones.

Unfortunately nonconvex functions do not have these properties, because we may for example have $f(x_k) - f_j^k < 0$. Due to this we have to generalize the linearization errors so that they retain the desirable properties of measures described above.

In what follows we shall use so-called *subgradient locality measures* of **Kiwiel** (1985) to generalize the linearization errors for nonconvex functions. For these locality measures we need a concept to measure the distance between the trial points and current iteration point.

Definition 3.1.3.2. At each iteration k we define the *distance measure* by

$$s_j^k := \begin{cases} \|x_j - y_j\| + \sum_{i=j}^{k-1} \|x_{i+1} - x_i\| & \text{for} \quad j = 1, \dots, k-1, \\ \|x_k - y_k\| & \text{for} \quad j = k, \end{cases}$$

the *aggregate distance measures* by $(s_f^1 := s_F^1 := 0)$

$$\begin{cases} \tilde{s}_f^k := \sum_{j \in J_f^k} \tilde{\lambda}_j^k s_j^k + \tilde{\lambda}_p^k s_f^k & \tilde{s}_F^k := \sum_{j \in J_F^k} \tilde{\mu}_j^k s_j^k + \tilde{\mu}_p^k s_F^k \\ s_f^{k+1} := \tilde{s}_f^k + \|x_{k+1} - x_k\| & s_F^{k+1} := \tilde{s}_F^k + \|x_{k+1} - x_k\| \end{cases}$$

and the *subgradient locality measures* by

$$\begin{cases} \beta_{f,j}^k := \max\{|\alpha_{f,j}^k|, \ \gamma_f(s_j^k)^2\} & \text{for all} \quad j \in J_f^k, \\ \beta_{f,p}^k := \max\{|\alpha_{f,p}^k|, \ \gamma_f(s_f^k)^2\}, \\ \tilde{\beta}_{f,p}^k := \max\{|\tilde{\alpha}_{f,p}^k|, \ \gamma_f(\tilde{s}_f^k)^2\}, \\ \beta_{F,j}^k := \max\{|\alpha_{F,j}^k|, \ \gamma_F(s_j^k)^2\} & \text{for all} \quad j \in J_F^k, \\ \beta_{F,p}^k := \max\{|\alpha_{F,p}^k|, \ \gamma_F(s_F^k)^2\}, \\ \tilde{\beta}_{F,p}^k := \max\{|\tilde{\alpha}_{F,p}^k|, \ \gamma_F(\tilde{s}_F^k)^2\}, \\ \tilde{\beta}_p^k := \lambda_f^k \tilde{\beta}_{f,p}^k + \mu_F^k \tilde{\beta}_{F,p}^k, \end{cases}$$

where $\gamma_f \geq 0$ and $\gamma_F \geq 0$ are the *distance measure parameters* ($\gamma_f = 0$ if f is convex and $\gamma_F = 0$ if F is convex).

Corollary 3.1.3.3. *For convex problem functions the subgradient locality measures and the linearization errors coincide.*

Proof. Follows from Theorem 3.1.3.1(viii) and the fact that for convex problem functions the distance measure parameters are zero. □

Remark 3.1.3.4. *Our simple idea to avoid the drawbacks caused by nonconvexity is to replace the linearization errors α by the subgradient locality measures β in our direction finding problems (ABP) and (ADP).*

3.2. Line Search

In this section we address the question of how to obtain the next iterate x_{k+1}. Suppose that at the k-th iteration we have found a solution (d_k, v_k) of the problem (ABP) with the subgradient locality measures. Although the direction d_k minimizes the polyhedral approximation $\hat{H}^k$ subject to the linear constraints, we have to remember that it is only an approximation to H. For this reason the direct choice $x_{k+1} := x_k + d_k$ is not necessarily the best one and hence we try to find a step size $t_k \in (0, 1]$ such that

$$t_k \approx \arg \min_{t \in (0,1]} \{f(x_k + td_k)\} \qquad \text{and} \qquad x_k + t_k d_k \in G.$$

Unfortunately also the choice $x_{k+1} := x_k + t_k d_k$ is not always justified. As mentioned before the direction d_k may even be a nondescent one or it may not yield a sufficiently large descent for f. In such cases the algorithm might take infinitely many steps without significantly reducing the objective value, which might impair convergence.

Due to these facts we have to modify this conventional line search. First we present some results which justify some ideas for the line search.

Theorem 3.2.1. *Let (d_k, v_k) be the solution of the problem (ABP) with the subgradient locality measures. Then*

(i) $\hat{f}^k(x_k + d_k) - f(x_k) \leq v_k \leq 0,$

(ii) $\hat{f}^k(x_k + td_k) \leq f(x_k) + t v_k \quad \text{for all} \quad t \in [0, 1].$

Proof. By applying the Corollary 3.1.2.4(j) to problem (ABP) with the subgradient locality measures we obtain

$$v_k \geq -\beta_{f,j}^k + (\xi_j^f)^{\mathrm{T}} d_k \geq f_j^k + (\xi_j^f)^{\mathrm{T}}(x_k + d_k - x_k) - f(x_k) \qquad \text{for all} \quad j \in J_f^k,$$

which implies by definition of the polyhedral approximation $\hat{f}^k(x)$ that

$$v_k \geq \hat{f}^k(x_k + d_k) - f(x_k).$$

On the other hand the vector $(d, v) := (0, 0)$ is a feasible solution of problem (ABP) with the subgradient locality measures. Then for the optimal (d_k, v_k) we have

$$v_k + \frac{u_k}{2} \|d_k\| \leq 0,$$

hence $v_k \leq 0$. Since $\hat{f}^k$ is convex, the assertion (i) and the fact $\hat{f}^k(x_k) = f(x_k)$ imply that

$$\hat{f}^k(x_k + td_k) \leq (1 - t)f(x_k) + t\hat{f}^k(x_k + d_k)$$
$$= f(x_k) + t(\hat{f}^k(x_k + d_k) - f(x_k))$$
$$\leq f(x_k) + tv_k \quad \text{for all} \quad t \in [0,1].$$

$\square$

We shall now describe a two-point line search, which will detect discontinuities in the gradient of f. We assume that

$$\begin{cases} m_L \in (0, \tfrac{1}{2}) \\ m_R \in (m_L, 1) \\ \bar{t} \in (0, 1] \\ \zeta := 1 - \tfrac{1}{2} \cdot \frac{1}{1 - m_L} \end{cases}$$

are fixed line search parameters. First we shall search for the largest number $t_L^k \in [0, 1]$ such that

    (a) $f(x_k + t_L^k d_k) \leq f(x_k) + m_L t_L^k v_k$,

    (b) $F(x_k + t_L^k d_k) \leq 0$,

    (c) $C(x_k + t_L^k d_k) \leq b$,

    (d) $t_L^k \geq \bar{t}$.

If such a parameter exists we take

$$\text{a long serious step}: \quad x_{k+1} := x_k + t_L^k d_k \quad \text{and} \quad y_{k+1} := x_{k+1}. \qquad (3.18)$$

Otherwise if requirements (a)–(c) hold but $0 < t_L^k < \bar{t}$ then we take

$$\text{a short serious step}: \quad x_{k+1} := x_k + t_L^k d_k \quad \text{and} \quad y_{k+1} := x_k + t_R^k d_k \qquad (3.19)$$

and if $t_L^k = 0$ we take

$$\text{a null step}: \quad x_{k+1} := x_k \quad \text{and} \quad y_{k+1} := x_k + t_R^k d_k \qquad (3.20)$$

where $t_R^k > t_L^k$ is such that

    (e) $-\beta_{f,k+1}^{k+1} + (\xi_{k+1}^f)^T d_k \geq m_R v_k$.

In long serious steps there occurs a significant decrease of the objective function. For this reason there is no need for detecting discontinuities in the gradient of f and so we set $\xi_{k+1}^f \in \partial f(x_{k+1})$. In short serious steps and null steps there exists discontinuity in the gradient of f. Then the requirement (e) ensures that x_k and y_{k+1} lie on the opposite sides of a discontinuity of the gradient and the new subgradient $\xi_{k+1}^f \in \partial f(y_{k+1})$ will force a significant modification of the next search direction finding problem. We shall now present a line search algorithm which finds step sizes t_L^k and t_R^k such that requirements (a)–(e) hold.

Algorithm 3.2.2 (Line search).

Step (i): Set $t_L^k := 0$ and $t := t_U := 1$.

Step (ii): If
$$\begin{cases} f(x_k + td_k) \leq f(x_k) + m_L t v_k \\ F(x_k + td_k) \leq 0 \\ C(x_k + td_k) \leq b \end{cases}$$

set $t_L^k := t$, otherwise set $t_U := t$.

Step (iii): If $t_L^k \geq \bar{t}$ set $t_R^k := t_L^k$ and STOP, otherwise calculate $\xi^f \in \partial f(x_k + td_k)$ and
$$\beta := \max\left\{ |f(x_k + t_L^k d_k) - f(x_k + td_k) + (t - t_L^k)(\xi^f)^{\mathrm{T}} d_k|, \right.$$
$$\left. \gamma_f(t - t_L^k)^2 \|d_k\|^2 \right\}.$$

Then if
$$-\beta + (\xi^f)^{\mathrm{T}} d_k \geq m_R v_k, \tag{3.21}$$

set $t_R^k := t$ and STOP.

Step (iv): If $t_L^k = 0$, then set
$$t := \max\left\{ \zeta \cdot t_U, \; \frac{\frac{1}{2} t_U^2 v_k}{t_U v_k + f(x_k) - f(x_k + t_U d_k)} \right\} \tag{3.22}$$

and if $t_L^k > 0$, then set $t := \frac{1}{2} \cdot (t_L^k + t_U)$

Step (v): Go to step *(ii)*.

We shall now show that this line search algorithm fulfills the line search requirements (a)–(e). However, we shall first analyze the step *(iv)* and present two lemmas. The first one will justify the choice of t in (3.22) and the second one will show that after each iteration the interval containing t is decreased.

Lemma 3.2.3. *Consider the quadratic function $g : \mathbf{R} \to \mathbf{R}$ that interpolates the function $t \mapsto f(x_k + td_k)$ in the sense that $g(0) = f(x_k)$, $g(t_U) = f(x_k + t_U d_k)$ and $g'(0) = v_k$, then*

$$\arg\min\{g(t) \mid t \in \mathbf{R}\} = \frac{\frac{1}{2}t_U^2 v_k}{t_U v_k + f(x_k) - f(x_k + t_U d_k)}. \tag{3.23}$$

Proof. The function g is in the form $g(t) = at^2 + bt + c$, where

$$\begin{cases} a = \dfrac{f(x_k + t_U d_k) - f(x_k) - t_U v_k}{t_U^2} \\ b = v_k \\ c = f(x_k) \end{cases}$$

and then we have

$$\arg\min\{g(t) \mid t \in \mathbf{R}\} = \frac{-b}{2a}.$$

$\square$

Lemma 3.2.4. *After step (iv) of Algorithm 3.2.2 we have*

$$t \in \left[t_L^k + \zeta(t_U - t_L^k), \ t_U - \zeta(t_U - t_L^k)\right]. \tag{3.24}$$

Proof. The statement is evident in the case when $t_L^k > 0$, since $\zeta \in (0, \frac{1}{2})$. Suppose now that $t_L^k = 0$, which implies by step (ii) that $t_U = t > 0$ and

$$f(x_k + t_U d_k) > f(x_k) + m_L t_U v_k. \tag{3.25}$$

By definition (3.22) $t \geq \zeta \cdot t_U$ and if $t > \zeta \cdot t_U$ then by (3.25) we have

$$\begin{aligned} t &= \frac{\frac{1}{2}t_U^2 v_k}{t_U v_k + f(x_k) - f(x_k + t_U d_k)} \\ &\leq \frac{\frac{1}{2}t_U^2 v_k}{t_U v_k + f(x_k) - f(x_k) - m_L t_U v_k} \\ &= \frac{\frac{1}{2}t_U}{1 - m_L} = (1 - \zeta)t_U, \end{aligned}$$

which implies that

$$t \in [\zeta t_U, (1 - \zeta)t_U] = \left[t_L^k + \zeta(t_U - t_L^k), \ t_U - \zeta(t_U - t_L^k)\right].$$

$\square$

Theorem 3.2.5. *The step sizes t_L^k and t_R^k generated by the Algorithm 3.2.2 fulfill the line search requirements (a)–(e).*

Proof. There are three different conditions for termination in step *(iii)*. First if $t_L^k > 0$, then the initialization $t_L^k = 0$ guarantees that in step *(ii)* we have had

$$\begin{cases} f(x_k + t_L^k d_k) \leq f(x_k) + m_L t_L^k v_k \\ F(x_k + t_L^k d_k) \leq 0 \\ C(x_k + t_L^k d_k) \leq b. \end{cases}$$

If in addition $t_L^k \geq \bar{t}$, then the requirements (a)–(d) are fulfilled and we take the long serious step by setting $t_R^k = t_L^k$.

Another termination situation occurs when the requirements (a)–(c) are valid, but $0 < t_L^k < \bar{t}$ and inequality in (3.21) holds. Then by setting $t_R^k = t$ we have

$$\begin{aligned} \beta &= \max \left\{ |f(x_{k+1}) - f(y_{k+1}) - (\xi_{k+1}^f)^{\mathrm{T}}(x_{k+1} - y_{k+1})|, \ \gamma_f \|x_{k+1} - y_{k+1}\|^2 \right\} \\ &= \max \left\{ |f(x_{k+1}) - f_{k+1}^{k+1}|, \ \gamma_f(s_{k+1}^{k+1}) \right\} = \beta_{f,k+1}^{k+1}. \end{aligned}$$

Then inequality (3.21) implies the requirement (e) and we take the short serious step.

Finally if $t_L^k = 0$, then the requirements (a)–(c) are evident and by setting $t_R^k = t$ as above the requirement (e) is fulfilled and we take a null step. $\square$

3.3. Stopping Criterion

For a smooth function a necessary condition for a local minimum is that the gradient must be zero at each local solution, and by continuity it becomes small as soon as we are close to an optimal point. Unfortunately this is no longer true if we replace the gradient by an arbitrary subgradient.

The necessary condition for x to be a local optimum of the improvement function H over the linear feasible set G_C is

$$0 \in \partial H(x; x) + \sum_{i \in I} \nu_i C_i. \tag{3.26}$$

However, an arbitrary subgradient in $\partial H(x; x)$ need not satisfy the condition (3.26).

For this reason we have to be much more cautious. Since we use subgradient aggregation we have a quite useful approximation to the gradient, namely the aggregate subgradient p_k. However the direct test:

$$\|p_k\| < \varepsilon \quad \text{for some} \quad \varepsilon > 0 \tag{3.27}$$

is too uncertain and may cause harm if the current piecewise linear approximation is too rough. Thus we must have some measure of the accuracy of the linearizations. One suitable candidate is the aggregate subgradient locality measure $\tilde{\beta}_p^k$, which approximates the accuracy of the current linearization: if the value of the measure is large, then the linearization is rough and if it is near zero and (3.27) holds the linearization is quite accurate and we can stop the algorithm.

Definition 3.3.1. The *stopping parameter* at iteration k is defined by

$$w_k := \tfrac{1}{2}\|p_k\|^2 + \tilde{\beta}_p^k. \tag{3.28}$$

Let $\varepsilon_s > 0$ be an accuracy parameter supplied by the user. Now our stopping criterion at iteration k reads as follows.

- If $w_k < \varepsilon_s$ then STOP.

The next result justifies this stopping criterion in convex case.

Theorem 3.3.2. *If the problem functions are convex, then at every iteration k we have*

(i) $p_k \in \partial_{w_k} \hat{H}_{z_k}(x_k) + \sum_{i \in I} \nu_i^k C_i$ and

(ii) $\|p_k\| \leq \sqrt{2w_k}$.

Proof. Assertion (i) follows from Theorem 3.1.3.1(viii) and the fact that

$$w_k = \tfrac{1}{2}\|p_k\|^2 + \tilde{\beta}_p^k \geq \tilde{\beta}_p^k = \tilde{\alpha}_p^k.$$

For assertion (ii) we have

$$\|p_k\| = \sqrt{\|p_k\|^2} \leq \sqrt{\|p_k\|^2 + 2\tilde{\beta}_p^k} = \sqrt{2w_k}.$$

$\square$

3.4. Weight Updating

The last but not least important open question is the choice of the weight u_k. The simplest strategy might be to keep it fixed $u_k \equiv u_{fix}$. This leads, however, to several difficulties. Due to assertions (f) and (g) of Theorem 3.1.1.3(iii) we observe the following.

(i) If u_{fix} is very large, we shall have small $|v_k|$ and $\|d_k\|$, almost all steps serious and slow descent.

(ii) If u_{fix} is very small, we shall have large $|v_k|$ and $\|d_k\|$, and each serious step will be followed by many null steps.

For these reasons we keep u_k as a variable and update it when necessary. In what follows we shall present a safeguarded quadratic interpolation technique by **Kiwiel (1990)** for updating u_k.

After a long serious step with $t_L^k = 1$ (a full step) we shall detect whether u_k is too large (cf. (i)). Due to the line search requirements we have $y_{k+1} = x_k + d_k$ and

$$f(y_{k+1}) \leq f(x_k) + m_L v_k$$

and if

$$f(y_{k+1}) \leq f(x_k) + m_R v_k \tag{3.29}$$

we decide that u_k should be decreased ($m_R \in (m_L, 1)$). We are using the following quadratic interpolation scheme

$$u_{k+1}^{int} := 2u_k\big(1 - [f(y_{k+1}) - f(x_k)]/v_k\big), \tag{3.30}$$

which is established by the following result.

Lemma 3.4.1. *Consider the problem*

$$\begin{cases} minimize & f(x) \\ subject\ to & x \in \mathbf{R}, \end{cases}$$

where $f : \mathbf{R} \to \mathbf{R}$ *is quadratic and strictly convex. Then we get the optimal solution at iteration 3 with updating* $u_k := u_k^{int}$.

Proof. The function f is of the form ($a > 0$)

$$f(x) = \tfrac{1}{2}ax^2 + bx + c$$

and at $k = 1$ we have $v_1 = \xi_1^T d_1 = -u_1 \|d_1\|^2$. Then

$$f(y_2) = f(x_1) + v_1 + \frac{a}{2}\|d_1\|^2,$$

which implies that

$$a = u_2^{int} = 2u_1\left(1 - [f(y_2) - f(x_1)]/v_1\right).$$

Now the update $u_2 := u_2^{int} = a$ gives the polyhedral approximation $\hat{f}^2$ which equals to f at x_1 and y_2 and thus

$$\hat{f}^2(x_2 + d) + \frac{u_2}{2}\|d\|^2 \tag{3.31}$$

equals to f, since $a = u_2$. We find d_2 by minimizing (3.31) which implies that $y_3 = x_2 + d_2$ is optimal. Further we have

$$f(y_3) = f(x_2) + \frac{v_2}{2}$$

and then the trial $t_L^2 = 1$ with $m_L \in (0, \frac{1}{2})$ leads to

$$f(y_3) \le f(x_2) + m_L t_L^2 v_2$$

which fulfills the line search requirements and we set $x_3 := y_3$. $\square$

For a general f due to (3.29) and (3.30) with $m_R \in (\frac{1}{2}, 1)$ we have

$$\begin{aligned}
u_{k+1}^{int} &= 2u_k\left(1 - [f(y_{k+1}) - f(x_k)]/v_k\right) \\
&\le 2u_k(1 - m_R) \\
&< u_k.
\end{aligned}$$

To avoid some numerical problems we safeguard our quadratic interpolation by letting

$$u_{k+1} := \max\{u_{k+1}^{int}, u_k/10, u_{min}\}, \tag{3.32}$$

where $u_{min} > 0$ is a small constant.

In the cases when condition (3.29) does not hold but we have some (say for example four) consecutive full steps, we also decide to decrease u_k by updating

$$u_{k+1} := \max\{u_k/2, u_{min}\}. \tag{3.33}$$

On the other hand after some consecutive null steps we use the test

$$\beta_{f,k+1}^{k+1} > \max \{\|p_k\| + \tilde{\beta}_p^k, -10v_k\} \tag{3.34}$$

for deciding that u_k should be increased. If (3.34) holds we update u_k by

$$u_{k+1} := \min \{u_{k+1}^{int}, 10u_k\}. \tag{3.35}$$

We shall now present a weight updating algorithm proposed in **Kiwiel** (1990). We denote by ε_v^k the *variation estimate* which corresponds to the size of $\|p_k\| + \tilde{\beta}_p^k$ and by i_u^k the step counter which counts the number of long serious steps with $t_L^k = 1$ and null steps since the latest change of u_k. These variables are initialized by

$$\varepsilon_v^1 := +\infty \qquad \text{and} \qquad i_u^1 := 0.$$

Algorithm 3.4.2 (Weight Updating).

Step (i): Set $u := u_k$.

Step (ii): If $t_L^k \in (0,1)$ STOP.

Step (iii): If $t_L^k = 0$ go to step *(v)*.

Step (iv): Set

$$\varepsilon_v^{k+1} := \max \{\varepsilon_v^k, -2v_k\}.$$

If $i_u^k > 0$ and

$$f(y_{k+1}) \leq f(x_k) + m_R v_k,$$

set $u := u_{k+1}^{int}$, otherwise, if $i_u^k > 3$ set $u := u_k/2$. Set

$$u_{k+1} := \max \{u, u_k/10, u_{min}\},$$
$$i_u^{k+1} := \max \{i_u^k + 1, 1\}.$$

If $u_{k+1} \neq u_k$ set $i_u^{k+1} := 1$. STOP.

Step (v): Set

$$\varepsilon_v^{k+1} := \min \{\varepsilon_v^k, \|p_k\| + \tilde{\beta}_p^k\}.$$

If $i_u^k < -3$ and

$$\beta_{f,k+1}^{k+1} > \max \{\varepsilon_v^{k+1}, -10v_k\},$$

set $u := u_{k+1}^{int}$. Set

$$u_{k+1} := \min \{u, 10u_k\},$$
$$i_u^{k+1} := \min \{i_u^k - 1, -1\}.$$

If $u_{k+1} \neq u_k$ set $i_u^{k+1} := -1$. STOP.

3.5. Algorithm

We are now ready to present the Proximal Bundle Method for nonconvex constrained optimization.

Algorithm 3.5.1 (Proximal Bundle).

Step 0: *(Initialization)*. Select a starting point $x_1 \in G$, a final accuracy tolerance $\varepsilon_s > 0$, the maximum number of stored subgradients $M_\xi \geq 2$, an initial weight $u_1 > 0$, a lower bound for weights $u_{min} > 0$ and line search parameters $m_L \in (0, \frac{1}{2})$, $m_R \in (m_L, 1)$ and $\bar{t} \in (0, 1]$. Choose the distance measure parameters $\gamma_f > 0$ and $\gamma_F > 0$ ($\gamma_f = 0$ if f is convex; $\gamma_F = 0$ if F is convex). Set the iteration counter $k := 1$ and the line search parameter $\zeta := 1 - \frac{1}{2} \cdot \frac{1}{1-m_L}$ and initialize the following variables:

$$\begin{cases} y_1 := x_1, & p_f^0 := \xi_1^f \in \partial f(y_1), & f_p^1 := f_1^1 := f(y_1), & s_f^1 := s_1^1 := 0, \\ J_1 := \{1\}, & p_F^0 := \xi_1^F \in \partial F(y_1), & F_p^1 := F_1^1 := F(y_1), & s_F^1 := 0. \end{cases}$$

Step 1: *(Direction finding)*. Find multipliers λ_p^k, μ_p^k, λ_j^k for $j \in J_f^k$, μ_j^k for $j \in J_F^k$ and ν_j^k for $i \in I$ by solving the following dual problem

$$\begin{cases} \text{minimize} & \dfrac{1}{2u_k} \| \sum_{j \in J_f^k} \lambda_j \xi_j^f + \lambda_p p_f^{k-1} + \sum_{j \in J_F^k} \mu_j \xi_j^F + \mu_p p_F^{k-1} + \sum_{i \in I} \nu_i C_i \|^2 \\ & + \sum_{j \in J_f^k} \lambda_j \beta_{f,j}^k + \lambda_p \beta_{f,p}^k + \sum_{j \in J_F^k} \mu_j \beta_{F,j}^k + \mu_p \beta_{F,p}^k + \sum_{i \in I} \nu_i \beta_{C,i}^k \\ \text{subject to} & \sum_{j \in J_k} \lambda_j + \lambda_p + \sum_{j \in J_F^k} \mu_j + \mu_p = 1 \\ \text{and} & \lambda_j, \lambda_p, \mu_j, \mu_p, \nu_i \geq 0, \end{cases}$$

where

$$\beta_{f,j}^k := \max \{|f(x_k) - f_j^k|, \gamma_f(s_j^k)^2\} \quad \text{for all} \quad j \in J_f^k,$$
$$\beta_{f,p}^k := \max \{|f(x_k) - f_p^k|, \gamma_f(s_f^k)^2\},$$
$$\beta_{F,j}^k := \max \{|-F_j^k|, \gamma_F(s_j^k)^2\} \quad \text{for all} \quad j \in J_F^k,$$
$$\beta_{F,p}^k := \max \{|-F_p^k|, \gamma_F(s_F^k)^2\} \quad \text{and}$$
$$\beta_{C,i}^k := -C_i^k \quad \text{for all} \quad i \in I.$$

Calculate multipliers λ_f^k, μ_F^k, $\tilde{\lambda}_p^k$, $\tilde{\mu}_p^k$, $\tilde{\lambda}_j^k$ and $\tilde{\mu}_j^k$ for $j \in J_f^k$ and $j \in J_F^k$ by (3.22) and set

$$(p_f^k, \tilde{f}_p^k, \tilde{s}_f^k) := \sum_{j \in J_f^k} \tilde{\lambda}_j^k(\xi_j^f, f_j^k, s_j^k) + \tilde{\lambda}_p^k(p_f^{k-1}, f_p^k, s_f^k),$$

$$(p_F^k, \tilde{F}_p^k, \tilde{s}_F^k) := \sum_{j \in J_F^k} \tilde{\mu}_j^k(\xi_j^F, F_j^k, s_j^k) + \tilde{\mu}_p^k(p_F^{k-1}, F_p^k, s_F^k),$$

$$p_k := \lambda_f^k p_f^k + \mu_F^k p_F^k + \sum_{i \in I} \nu_i^k C_i,$$

$$\tilde{\beta}_{f,p}^k := \max\{|f(x_k) - \tilde{f}_p^k|, \ \gamma_f(\tilde{s}_f^k)^2\},$$

$$\tilde{\beta}_{F,p}^k := \max\{|-\tilde{F}_p^k|, \ \gamma_F(\tilde{s}_F^k)^2\} \quad \text{and}$$

$$\tilde{\beta}_p^k := \lambda_f^k \tilde{\beta}_{f,p}^k + \mu_F^k \tilde{\beta}_{F,p}^k.$$

Set $d_k := -\frac{1}{u_k}p_k$.

Step 2: (Stopping criterion). Set

$$w_k := \tfrac{1}{2}\|p_k\|^2 + \tilde{\beta}_p^k.$$

If $w_k \leq \varepsilon_s$ then STOP.

Step 3: (Line search). By line search Algorithm 3.2.2 find step sizes $t_L^k \in [0,1]$ and $t_R^k \in [t_L^k, 1]$. Set

$$x_{k+1} := x_k + t_L^k d_k \quad \text{and} \quad y_{k+1} := x_k + t_R^k d_k.$$

Step 4: (Linearization updating). Calculate the linearization values

$$f_j^{k+1} := f_j^k + t_L^k(\xi_j^f)^{\mathrm{T}} d_k, \quad \text{for} \quad j \in J_f^k,$$

$$F_j^{k+1} := F_j^k + t_L^k(\xi_j^F)^{\mathrm{T}} d_k, \quad \text{for} \quad j \in J_F^k,$$

$$s_j^{k+1} := s_j^k + t_L^k\|d_k\|, \quad \text{for} \quad j \in J_f^k \cup J_F^k,$$

$$f_p^{k+1} := \tilde{f}_p^k + t_L^k(p_f^k)^{\mathrm{T}} d_k,$$

$$F_p^{k+1} := \tilde{F}_p^k + t_L^k(p_F^k)^{\mathrm{T}} d_k,$$

$$s_f^{k+1} := \tilde{s}_f^k + t_L^k\|d_k\|,$$

$$s_F^{k+1} := \tilde{s}_F^k + t_L^k\|d_k\|.$$

Evaluate $\xi^f_{k+1} \in \partial f(y_{k+1})$ and $\xi^F_{k+1} \in \partial F(y_{k+1})$ and set

$$f^{k+1}_{k+1} := f(y_{k+1}) + (t^k_L - t^k_R)(\xi^f_{k+1})^T d_k,$$
$$F^{k+1}_{k+1} := F(y_{k+1}) + (t^k_L - t^k_R)(\xi^F_{k+1})^T d_k,$$
$$s^{k+1}_{k+1} := (t^k_R - t^k_L)\|d_k\|.$$

Step 5: (Weight updating). Select u_{k+1} by the weight updating Algorithm 3.4.2.

Step 6: (Updating). Set $J^{k+1}_f := J^k_f \cup \{k+1\}$ and $J^{k+1}_F := J^k_F \cup \{k+1\}$. If card $J^{k+1}_f > M_\xi$, then $J^{k+1}_f := J^{k+1}_f \setminus \{\min j \mid j \in J^{k+1}_f\}$. If card $J^{k+1}_F > M_\xi$, then $J^{k+1}_F := J^{k+1}_F \setminus \{\min j \mid j \in J^{k+1}_F\}$. Increase k by 1 and go to Step 1.

Remark 3.5.2. *A convergence proof for the above algorithm may be developed as in* **Kiwiel** *(1985) or in* **Schramm and Zowe** *(1990) by assuming, that f is weakly semismooth, i.e. the directional derivative $f'(x; d)$ exists for all $x \in \mathbf{R}^n$ and $d \in \mathbf{R}^n$, and $f'(x; d) = \lim_{t \downarrow 0} \xi^T d$, where $\xi \in \partial f(x + td)$.*

For further study of the methods in nonsmooth optimization we refer to **Allen, Helgason, Kennigton and Shetty** (1987), **Balinski and Wolfe** (1975), **Clarke** (1989, 1983), **Demyanov and Dixon** (1986), **Demyanov, Lemaréchal and Zowe** (1986), **Demyanov and Pallaschke** (1984), **Donno and Pesamosca** (1990), **Fletcher and Sainz de la Maza** (1989), **Gaudioso and Monaco** (1988, 1991), **Ha** (1990), **Kim, Koh and Ahn** (1987), **Kiwiel** (1985, 1986a, 1986b, 1987a, 1987b, 1989a, 1989b), **Kurzhanski, Neumann and Pallaschke** (1987), **Lemaréchal** (1976, 1986a, 1989b), **Lemaréchal and Strodiot** (1984), **Mifflin** (1977, 1984), **Mifflin and Strodiot** (1989), **Mäkelä** (1990a), **Outrata** (1983, 1986, 1987, 1988, 1990), **Panier** (1987), **Sachs** (1983), **Shor** (1983, 1985), **Schramm** (1989), **Schramm and Zowe** (1990), **Zowe** (1985, 1987, 1988), **Uryasev** (1991) and **Yuan** (1985).

Chapter 4

Numerical Experiments

The Proximal Bundle (PB) method described in the previous chapter was implemented in Fortran 77. To show the reliability of our code PB, some numerical experiments with classical nonsmooth test problems from the literature will now be reported. All calculations are made in double precision on a VAX 8650 computer with relative accuracy $\varepsilon_M \approx 2.776 \cdot 10^{-17}$. To solve the quadratic direction finding problem in step 1 of Algorithm 3.5.1 we employ the subroutines QPDF2 and QPDF4 of Kiwiel (1986a) for unconstrained problems and those with simple bounds, respectively. For linear and nonlinear constraints we use the subroutine E04NAF from the NAG-subroutine library. We shall compare our numerical results with two other nonsmooth optimization methods. The following codes were tested

Table 4.1

Code	Author	Method
M1FC1	C.Lemaréchal	ε-steepest descent
BT	H.Schramm	Bundle Trust
PB	M.M.Mäkelä	Proximal Bundle

4.1. Test Examples

First we give some standard examples from the literature. To save space, we shall only sketch the structure of some problems, for more details see for example **Lemaréchal and Mifflin (1977)**, **Shor (1985)**, **Schramm (1989)**, **Kiwiel (1990)** and references therein. For simplicity we tested only unconstrained examples.

Rosenbrock
Dimension   2,
Objective function $f(x) = 100(x_2 - x_1^2)^2 + (1 - x_1)^2$,
Optimum point $x^* = (1, 1)$,
Optimum value $f(x^*) = 0$,
Starting point $x^1 = (-1.2, 1)$.

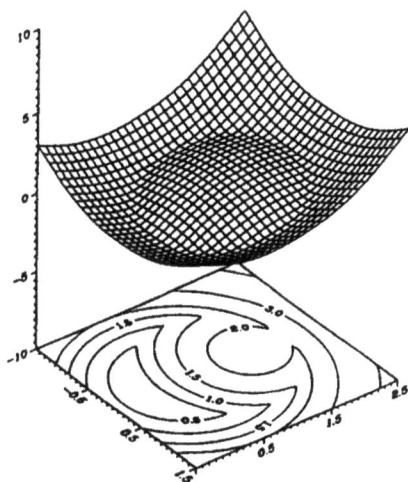

Figure 4.1. Crescent

Crescent (see Figure 4.1)
Dimension   2,
Objective function $f(x) = \max\{x_1^2 + (x_2 - 1)^2 + x_2 - 1,$
$$-x_1^2 - (x_2 - 1)^2 + x_2 + 1\},$$
Optimum point $x^* = (0, 0)$,
Optimum value $f(x^*) = 0$,
Starting point $x^1 = (-1.5, 2)$.

CB2 (Charalambous/Bandler)
Dimension   2,
Objective function $f(x) = \max\{x_1^2 + x_2^4, (2 - x_1)^2 + (2 - x_2)^2, 2e^{-x_1 + x_2}\}$,
Optimum point $x^* = (1.139286, 0.899365)$,
Optimum value $f(x^*) = 1.952225$,
Starting point $x^1 = (1, -0.1)$.

CB3 (Charalambous/Bandler)
Dimension   2,
Objective function $f(x) = \max\{x_1^4 + x_2^2, (2 - x_1)^2 + (2 - x_2)^2, 2e^{-x_1 + x_2}\}$,
Optimum point $x^* = (1, 1)$,
Optimum value $f(x^*) = 2$,
Starting point $x^1 = (2, 2)$.

DEM (Demyanov/Malozemov)

Dimension	2,
Objective function	$f(x) = \max\{5x_1 + x_2, -5x_1 + x_2, x_1^2 + x_2^2 + 4x_2\}$,
Optimum point	$x^* = (0, -3)$,
Optimum value	$f(x^*) = -3$,
Starting point	$x^1 = (1, 1)$.

QL

Dimension	2,
Objective function	$f(x) = \max_{1 \leq i \leq 3} f_i(x)$,
where	$f_1(x) = x_1^2 + x_2^2$,
	$f_2(x) = x_1^2 + x_2^2 + 10(-4x_1 - x_2 + 4)$,
	$f_3(x) = x_1^2 + x_2^2 + 10(-x_1 - 2x_2 + 6)$,
Optimum point	$x^* = (1.2, 2.4)$,
Optimum value	$f(x^*) = 7.2$,
Starting point	$x^1 = (-1, 5)$.

LQ

Dimension	2,
Objective function	$f(x) = \max\{-x_1 - x_2, -x_1 - x_2 + (x_1^2 + x_2^2 - 1)\}$,
Optimum point	$x^* = (\frac{1}{\sqrt{2}}, \frac{1}{\sqrt{2}})$
Optimum value	$f(x^*) = -\sqrt{2}$,
Starting point	$x^1 = (-0.5, -0.5)$.

Mifflin1

Dimension	2,
Objective function	$f(x) = -x_1 + 20 \max\{x_1^2 + x_2^2 - 1, 0\}$,
Optimum point	$x^* = (1, 0)$,
Optimum value	$f(x^*) = -1$,
Starting point	$x^1 = (0.8, 0.6)$.

Mifflin2

Dimension	2,
Objective function	$f(x) = -x_1 + 2(x_1^2 + x_2^2 - 1) + 1.75|x_1^2 + x_2^2 - 1|$,
Optimum point	$x^* = (1, 0)$,
Optimum value	$f(x^*) = -1$,
Starting point	$x^1 = (-1, -1)$.

Rosen

Dimension | 4,
Objective function | $f(x) = \max\{f_1(x), f_1(x) + 10f_2(x), f_1(x) + 10f_3(x),$
| $f_1(x) + 10f_4(x)\}$,
where | $f_1(x) = x_1^2 + x_2^2 + 2x_3^2 + x_4^2 - 5x_1 - 5x_2 - 21x_3 + 7x_4,$
| $f_2(x) = x_1^2 + x_2^2 + x_3^2 + x_4^2 + x_1 - x_2 + x_3 - x_4 - 8,$
| $f_3(x) = x_1^2 + 2x_2^2 + x_3^2 + 2x_4^2 - x_1 - x_4 - 10,$
| $f_4(x) = x_1^2 + x_2^2 + x_3^2 + 2x_1 - x_2 - x_4 - 5,$
Optimum point | $x^* = (0, 1, 2, -1),$
Optimum value | $f(x^*) = -44,$
Starting point | $x^1 = (0, 0, 0, 0).$

Shor

Dimension | 5,
Objective function | $f(x) = \max_{1 \le i \le 10}\{b_i \sum_{j=1}^{5}(x_j - a_{ij})^2\},$
Optimum value | $f(x^*) = 22.60016,$
Starting point | $x^1 = (0, 0, 0, 0, 1).$

Maxquad

Dimension | 10,
Objective function | $f(x) = \max_{1 \le i \le 5}\{x^T A_i x - b_i^T x\},$
Optimum value | $f(x^*) = -0.8414084,$
Starting point 1 | $x^1 = (1, \ldots, 1),$
Starting point 2 | $x^1 = (0, \ldots, 0).$

Maxq

Dimension | 20,
Objective function | $f(x) = \max_{1 \le i \le 20} x_i^2,$
Optimum value | $f(x^*) = 0,$
Starting point | $x_i^1 = i, \quad i = 1, \ldots, 10,$
| $x_i^1 = -i, \quad i = 11, \ldots, 20.$

Maxl

Dimension | 20,
Objective function | $f(x) = \max_{1 \le i \le 20} |x_i|,$
Optimum value | $f(x^*) = 0,$
Starting point | $x_i^1 = i, \quad i = 1, \ldots, 10,$
| $x_i^1 = -i, \quad i = 11, \ldots, 20.$

TR48

Dimension	48,
Objective function	$f(x) = \sum_{j=1}^{48} d_j \max_{1 \le i \le 48} \{x_i - a_i j\} - \sum_{i=1}^{48} s_i x_i)$,
Optimum value	$f(x^*) = -638565$,
Starting point	$x_i^1 = 0, \quad i = 1, \dots, 48$.

Goffin

Dimension	50,
Objective function	$f(x) = 50 \max_{1 \le i \le 50} x_i - \sum_{i=1}^{50} x_i)$,
Optimum value	$f(x^*) = 0$,
Starting point	$x_i^1 = i - 25.5, \quad i = 1, \dots, 50$.

4.2. Numerical Results

Our results are summarized in Table 4.2. The comparison is made with those obtained in **Schramm** (1989). Similar test problems can also be found in **Kiwiel** (1990). The following abbreviations will be used:

$$
\begin{array}{rcl}
\textbf{it} & - & \text{the number of iterations} \\
\textbf{nf} & - & \text{the number of function and subgradient calls} \\
f^* & - & \text{the final value of the objective function.}
\end{array}
\qquad (4.1)
$$

The parameters had the following standard values: $m_L = 0.01$, $m_R = 0.5$, $\bar{t} = 0.01$, $u_{min} = 1.0 \cdot 10^{-10}$ and $\varepsilon_s = 1.0 \cdot 10^{-6}$. The first weight u_1 was chosen by $u_1 := \|\xi_1^f\|$. In nonconvex problems Rosenbrock and Crescent the convexity measure γ_f was chosen to be 0.3 and 1.0, respectively. To make the results compatible with those in **Schramm** (1989) the following values for upper limit of stored subgradients were used: $M_\xi = 5$ (CB2, CB3, DEM, QL and LQ), $M_\xi = 10$ (Rosenbrock, Crescent, Mifflin1, Mifflin2, Rosen, Shor and Maxquad) and $M_\xi = 200$ (Maxq, Maxl, TR48 and Goffin).

Table 4.2

Algorithm	M1FC1			BT			PB		
Problem	it	nf	f^*	it	nf	f^*	it	nf	f^*
Rosenbrock	70	121	$2.4 \cdot 10^{-7}$	79	88	$1.3 \cdot 10^{-12}$	42	43	$1.4 \cdot 10^{-8}$
Crescent	31	93	$2.2 \cdot 10^{-6}$	24	27	$9.4 \cdot 10^{-7}$	32	33	$5.4 \cdot 10^{-7}$
CB2	11	31	1.952253	13	16	1.952225	15	16	1.952225
CB3	12	44	2.001415	13	21	2.000000	15	16	2.000000
DEM	10	33	-3.000000	9	13	-3.000000	7	8	-3.000000
QL	12	30	7.200018	12	17	7.200009	17	18	7.200001
LQ	16	52	-1.141420	10	11	-1.414214	14	15	-1.414213
Mifflin1	143	281	-0.999967	49	74	-1.000000	22	23	-1.000000
Mifflin2	30	71	-0.999993	6	13	-1.000000	16	17	-1.000000
Rosen	22	61	-43.99998	22	32	-43.99998	40	41	-43.99999
Shor	21	69	22.60018	29	30	22.60016	26	27	22.60016
Maxquad1	29	69	-0.841359	45	56	-0.841408	41	42	-0.841408
Maxquad2	20	54	-0.841359	45	49	-0.841408	33	34	-0.841408
Maxq	144	207	0.000000	125	128	0.000000	158	159	0.000000
Maxl	138	213	0.000000	74	84	0.000000	34	35	0.000000
TR48	163	284	-633625.5	165	179	-638565.0	152	153	-638565.0
Goffin	72	194	46.2	50	53	0.000000	51	52	0.000000

Remark 4.2.1. *As a conclusion from our limited numerical experiments we may state that our method seems to work at least as efficiently as the others; in most examples PB required less function evaluations than M1FC1 and BT.*

For comparable numerical tests see for example **Schramm** (1989), **Schramm and Zowe** (1990), **Lemaréchal** (1982), and **Kiwiel** (1990).

Part III

Nonsmooth Optimal Control

Chapter 1

Introduction

The improvment of industrial processes as well as a new generation of CAD systems (such as computer aided structural design) lead us to use methods of optimal control. In optimal control problems the objective is to optimize (minimize or maximize) certain criteria involving the solution of the state system (partial differential equation for example) with respect to coefficient or the right hand side of the state system (distributed control problems), boundary values (boundary control problems) or with respect to the domain of definition of the state system (optimal shape design). When the discretization of the state system has been done (typically by finite element method in space and finite difference method in time) we are in a position to apply methods of optimization. In practice we often meet optimal control problems which lead to nonsmooth optimization.

The main aim of this part is to apply nonsmooth optimization for solving various optimal control problems. The material of this part is organized as follows.

Chapter 2 contains some basic formulae for sensitivity analysis as well as preliminaries from functional analysis.

Chapter 3 deals with optimal control problems where the constraints are both in state and control. We begin the study with an abstract approach and then give examples to demonstrate in detail the applications of nonsmooth optimization to solve the model problems. We shall compare the results with those obtained by smooth optimization technique in **Haslinger and Neittaanmäki** (1988). There are several monographs on optimal control theory for partial differential equations and on topics related to especially smooth optimization: **Ahmed** (1988), **Alekseev, Tikhomirov and Fomin** (1987), **Fiacco** (1983), **Lewis** (1986), **Lions** (1971, 1972, 1981), **Lurie** (1988) and **Teo and Wu** (1984).

Chapter 4 is devoted to optimal shape design problems. This class of problems is a central part of modern CAD systems. Namely, the primary problem often facing designers of structural systems is to determine the shape of the structure.

During the last 20 years, optimization theory has been widely developed for optimal design of many structural and mechanical systems. Optimal shape design problems have been studied quite extensively (see **Banichuk** (1990), **Haug and Céa** (1981) **Haslinger and Neittaanmäki** (1988), **Pironneau** (1984), **Mota Soares** (1987) and references therein). In practice we often meet problems, the behaviour of which, is described by variational inequalities. The fields of applications of variational inequalities are steadily growing (see **Barbu** (1984), **Duvaut and Lions** (1976), **Elliot and Ockendon** (1982), **Hlaváček, Haslinger, Nečas and Lovíšek** (1988), **Kikuchi and Oden** (1988), **Moreau, Panagiotopoulos and Strang** (1988), **Panagiotopoulos** (1985) and **Rodrigues** (1987)). The main aim of Chapter 3 is to investigate optimal shape design problems with state systems governed by variational inequalities. We begin the chapter with an abstract setting and study in detail concrete problems with unilateral boundary conditions (Dirichlet-Signorini boundary conditions). Moreover we consider the problem in finding the optimal covering for an obstacle. These problems were considered in **Haslinger and Neittaanmäki** (1988) by using regularization (exterior penalty method) and methods of smooth optimization.

The study of Chapter 5 is motivated by a real world problem: the steel continuous casting problem. In that problem, the state system (i.e. the temperature of the steel) should satisfy, as well as possible, some constraints that ensure a high quality of the steel.

Chapter 2

Preliminaries

2.1. Basics for Sensitivity Analysis

The discretization of optimal control problems often leads to optimization problems of the type

(P)
$$\underset{x \in X}{\text{minimize}} \ f(x) = h(x) + E(x, y(x))$$

subject to $y = y(x)$ satisfying a quadratic programming problem (the state system)

(S)
$$\underset{y \in \widetilde{K}}{\text{minimize}} \left\{ \tfrac{1}{2} y^{\mathrm{T}} A(x) y - G(x)^{\mathrm{T}} y \right\}.$$

Here $f \colon \mathbf{R}^n \to \mathbf{R}$, $h \colon \mathbf{R}^n \to \mathbf{R}$, $E \colon \mathbf{R}^n \times \mathbf{R}^m \to \mathbf{R}$, $y \colon \mathbf{R}^n \to \mathbf{R}^m$, $A \colon \mathbf{R}^n \to \mathbf{R}^{m \times m}$ and $G \colon \mathbf{R}^m \to \mathbf{R}^m$. Moreover,

$$X = \{ x \in \mathbf{R}^n \mid Cx \le b, \ x_{\min} \le x \le x_{\max} \}$$

and

$$\widetilde{K} = \{ y \in \mathbf{R}^m \mid Dy \le c, \ y_{\min} \le y \le y_{\max} \} \quad \text{or} \quad \widetilde{K} = \mathbf{R}^m,$$

where C is a $k \times n$ matrix and b is a k-vector and respectively D is a $l \times m$ matrix and c is a l-vector. We suppose that A is positive definite. In the case $\widetilde{K} = \mathbf{R}^m$ state problem (S) reduces to the state equation

(S′)
$$A(x)y = G(x).$$

In this case the mapping $x \mapsto y(x)$ is usually smooth but in general the function $J(x) = E(x, y(x))$ is nonsmooth because E may be (and often is) nonsmooth. In the case $\widetilde{K} \ne \mathbf{R}^m$, also the mapping $x \mapsto y(x)$ is only locally Lipschitz continuous and this causes that problem (P) is a nonsmooth optimization problem even if E would be smooth. Note that we could regard the variables x and y together as one vector (x, y) and formulate the system (S) as a constraint for problem (P), especially in the case where $\widetilde{K} = \mathbf{R}^m$. Then the problem would be differentiable. However, in practical applications the number of constraints m (the dimension of the state variable y) is usually very large and, due to the FEM-approximations, the problem (S) has a sparse nature. For these reasons it has proved to be advantageous to handle the problem (P) with the state system (S) as defined above.

149

To be able to apply nonsmooth optimization methods to problem (P) we have to calculate at least one subgradient from $\partial f(x)$ at each point $x \in \mathbf{R}^n$. This can be done by means of the following result.

Theorem 2.1. *Let $f\colon \mathbf{R}^n \to \mathbf{R}$ be defined by*

$$f(x) = h(x) + E(x, y(x)),$$

where the mapping $y\colon \mathbf{R}^n \to \mathbf{R}^m$ is locally Lipschitz and the functions $h\colon \mathbf{R}^n \to \mathbf{R}$ and $E\colon \mathbf{R}^n \times \mathbf{R}^m \to \mathbf{R}$ are continuously differentiable. If $\xi_y(x) \in \partial y(x)$, then

$$\nabla_x h(x) + \nabla_x E(x, y(x)) + \xi_y(x)^{\mathrm{T}}[\nabla_y E(x, y(x))] \in \partial f(x).$$

Proof. Follows directly from the classical chain rule and Corollary I 3.4.3. □

Theorem 2.2. *Let $f\colon \mathbf{R}^n \to \mathbf{R}$ be defined by*

$$f(x) = h(x) + E(y(x)),$$

where the mapping $y\colon \mathbf{R}^n \to \mathbf{R}^m$ and the function $h\colon \mathbf{R}^n \to \mathbf{R}$ are continuously differentiable, and $E\colon \mathbf{R}^m \to \mathbf{R}$ is convex. If $\xi_E(y(x)) \in \partial_y E(y(x))$, then

$$\nabla_x h(x) + \nabla_x y(x)^{\mathrm{T}} \xi_E(y(x)) \in \partial f(x).$$

Proof. Follows from Corollary I 3.4.3. □

However, the computation of a representative from the generalized Jacobian $\partial y(x)$ is normally highly complicated and available results concern only special cases (see **Haslinger and Roubiček** (1986), **Outrata** (1990)).

2.2. Preliminaries from Functional Analysis

Let X be a normed linear space equipped with the norm $\|\cdot\|_X$. A set $M \subset X$ is called *compact* if it is possible from an arbitrary sequence $\{u_n\}_{n=1}^{\infty}$ of elements of M to select a subsequence $\{u_{n_k}\}_{k=1}^{\infty}$ which converges to some element $u_0 \in M$, i.e.

$$\lim_{k \to \infty} \|u_{n_k} - u_0\|_X = 0 \,.$$

Furthermore, $M \subset X$ is called *weakly compact* if from an arbitrary sequence $\{u_n\}_{n=1}^{\infty}$ of elements of M it is possible to select a subsequence $\{u_{n_k}\}_{k=1}^{\infty}$ which converges weakly to an element of the set M, i.e. there exists $u_0 \in M$ such that $u_{n_k} \rightharpoonup u_0$. The term "weak convergence" means that for every continuous linear functional $f \in X'$ (dual space of X) we have

$$\lim_{n \to \infty} \langle f, u_n \rangle = \langle f, u_0 \rangle .$$

Lemma 2.3. *Let $J : M \to \mathbf{R}$ be a continuous functional on a non-empty compact subset M of the normed linear space X. Then there exists $u_0 \in M$ such that*

$$J(u_0) = \min_{u \in M} J(u) ,$$

i.e. $J(u_0) \le J(u)$ for every $u \in M$.

As the compact sets are too "small" when X is finite-dimensional, it is useful to re-formulate Lemma 2.3 in terms of the weak topology. Also, the condition of continuity of the functional J is unnecessarily strong. It suffices to assume that the functional J is *weakly lower semicontinuous* on the set M, i.e. that for every $u_0 \in M$ and for any sequence $\{u_n\}_{n=1}^{\infty}$ of elements of the set M such that $u_n \rightharpoonup u_0$ we have

$$J(u_0) \le \liminf_{n \to \infty} J(u_n) .$$

In the sequel, we will *assume* that the normed linear space X *is complete*, i.e. it is a Banach space.

Lemma 2.4. *Let M be a non-empty weakly compact subset of the Banach space X. Let J be a weakly lower semicontinuous functional on M. Then*

(a) $\inf\limits_{u \in M} J(u) > -\infty$;
(b) *there exists at least one $u_0 \in M$ such that*

$$J(u_0) = \min_{u \in M} J(u) .$$

Proof. See **Ekeland, Temam** (1976), for example. □

We can modify Lemma 2.3 in the following way.

Lemma 2.5. *Let M be a non-empty, convex, closed, and bounded subset of a reflexive Banach space X. Let J be a weakly lower semicontinuous functional on M. Then*

(a) $\inf\limits_{u \in M} J(u) > -\infty$;

(b) *there exists at least one $u_0 \in M$ such that*

$$J(u_0) = \min_{u \in M} J(u) \ .$$

Lebesgue and Sobolev spaces

Let p be a real number, $p \geq 1$ and let Ω be a domain in $\mathbf{R}^n$. By the symbol $L^p(\Omega)$ we denote the set of all (classes of) measurable functions in Ω for which the Lebesgue integral

$$\int_\Omega |u|^p \, d\Omega$$

is finite. The expression

$$\|u\|_{L^p(\Omega)} = \left[\int_\Omega |u|^p \, d\Omega \right]^{1/p}$$

defines a norm on $L^p(\Omega)$. $L^p(\Omega)$ with this norm is a Banach space.

By the symbol $L^\infty(\Omega)$ we denote the set of all measurable functions in Ω for which the number

$$\|u\|_{L^\infty(\Omega)} = \inf_{\substack{\text{meas } M = 0 \\ M \subset \Omega}} \sup_{s \in \Omega \setminus M} |u(s)| \equiv \operatorname*{ess\,sup}_{s \in \Omega} |u(s)| \tag{2.1}$$

is finite.

Formula (2.1) defines a norm on $L^\infty(\Omega)$ and $L^\infty(\Omega)$ is a Banach space with respect to this norm.

Let $\Omega \subset \mathbf{R}^n$ be a domain and let $k \in \mathbf{Z}_+$. The Sobolev space $W^{k,p}(\Omega)$, $p \geq 1$ is the subspace of $L^p(\Omega)$ of functions u for which there exist $\omega_\alpha \in L^p(\Omega)$, $\alpha = (\alpha_1, \alpha_2, \ldots, \alpha_n)$, $\alpha_i \in \mathbf{Z}_+$ for any α such that $0 \leq |\alpha| \leq k$, where $|\alpha| = \alpha_1 + \alpha_2 + \cdots + \alpha_n$, such that

$$\int_\Omega D^\alpha \phi u \, d\Omega = (-1)^{|\alpha|} \int_\Omega \phi \omega_\alpha \, d\Omega \qquad \forall \phi \in \mathcal{D}(\Omega) \ .$$

The function ω_α will be called the α-th distributional derivative of u and will be denoted by $D^\alpha u$ in what follows. For $1 \leq p < \infty$ denote

$$\|u\|_{k,p,\Omega} = \left(\sum_{0 \leq |\alpha| \leq k} \int_\Omega |D^\alpha u|^p \, d\Omega \right)^{1/p} \qquad (2.2)$$

and for $p = \infty$

$$\|u\|_{k,\infty,\Omega} = \sum_{0 \leq |\alpha| \leq k} \operatorname*{ess\,sup}_{s \in \Omega} |D^\alpha u(s)| . \qquad (2.3)$$

It is easy to see that $W^{k,p}(\Omega)$ is a Banach space equipped with norms (2.2) and (2.3), respectively.

If $k = 0$, then $W^{k,p}(\Omega)$ coincides with $L^p(\Omega)$ $(1 \leq p \leq \infty)$ and

$$\|u\|_{0,p,\Omega} = \|u\|_{L^p(\Omega)} .$$

Expressions

$$|u|_{k,p,\Omega} = \left(\sum_{|\alpha|=k} \int_\Omega |D^\alpha u|^p \, d\Omega \right)^{1/p} , \qquad 1 \leq p < \infty ,$$

$$|u|_{k,\infty,\Omega} = \sum_{|\alpha|=k} \operatorname*{ess\,sup}_{s \in \Omega} |D^\alpha u(s)|$$

are seminorms in $W^{k,p}(\Omega)$, $W^{k,\infty}(\Omega)$, respectively. We define

$$W_0^{k,p}(\Omega) = \{ u \in W^{k,p}(\Omega) \mid D^\alpha u = 0 \quad \text{a.e. on } \partial\Omega, \ |\alpha| \leq k-1 \}.$$

The case $p = 2$ is special, since $W^{k,p}(\Omega)$ will be a Hilbert space equipped with the scalar product

$$(u,v)_{k,\Omega} = \sum_{0 \leq |\alpha| \leq k} \int_\Omega D^\alpha u D^\alpha v \, ds$$

and we shall write

$$H^k(\Omega) = W^{k,2}(\Omega) \quad \text{and} \quad H_0^k(\Omega) = W_0^{k,2}(\Omega) .$$

Instead of $\| \cdot \|_{k,2,\Omega}$, $| \cdot |_{k,2,\Omega}$ we shall simply write $\| \cdot \|_{k,\Omega}$, $| \cdot |_{k,\Omega}$, respectively.

For further study of the material introduced in this chapter we refer to books **Adams** (1975), **Aubin** (1972), **Aubin and Ekeland** (1984), **Ekeland and Teman** (1976), **Fiacco** (1983), **Gill, Murray and Wright** (1981), **Glowinski** (1984) and **Nečas** (1967).

Chapter 3

Distributed Parameter Control Problems

3.1. Introduction

In this chapter we shall analyse the general situation where the state relation is given by equations or by inequalities and the control variable appears both in the coefficients and on the right hand side. Moreover, we impose additional constraints upon the state of the system. Such constraints are often of technological nature. We shall begin the study with an abstract setting of the problem and then we shall give several practical examples which fit into this setting. The abstract approach is based on the article **Haslinger, Neittaanmäki and Tiba (1988)**.

State constrained optimal control and optimal shape design problems have been studied by many authors during recent years, see for example **Barbu (1984)**, **Banichuk (1990)**, **Haslinger and Neittaanmäki (1988)**, **Tiba (1990)** and references therein.

3.2. Setting of the Problem. Existence of the Solution

Let V, U be two Banach spaces, $U_{ad} \subset U$, $K \subset V$ be convex, closed and nonempty subsets, $g \in V'$ (dual space of V), $B: U \to V'$ a linear continuous mapping and $\varphi: V \to (-\infty, +\infty]$ a convex, lower semicontinuous, proper function (i.e. for each $y \in V$ we have $\liminf_{v \to y} \varphi(v) \geq \varphi(y)$ and φ is not identically $+\infty$). The norms in V', U will be denoted by $\| \ \|_{V'}$ and $\| \ \|_U$, respectively. We shall deal with the optimization problem

(P)
$$\operatorname*{minimize}_{u \in U_{ad}} J(u, y),$$

where $u \in U_{ad}$ and $y \in V$ are related by the state problem

(S(u))
$$A(u)y + \partial\varphi(y) \ni Bu + g$$

with a state constraint

$$y \in K.$$

Here $J: U \times V \to \mathbf{R}$ is a cost functional, $A(u): V \to V'$ is a linear, continuous mapping for each $u \in U_{ad}$ and $\partial\varphi(y)$ denotes the subdifferential of φ at y.

Remark 3.1. If $\varphi \equiv 0$ then $(S(u))$ is equivalent to the equation

$$(S(u)') \qquad\qquad A(u)y = Bu + g.$$

If φ is the indicator function of a closed convex subset $\tilde{K} \subset V$, then $(S(u))$ transforms into the variational inequality

$$(S(u)'') \qquad \begin{cases} \text{Find } y \in \tilde{K} \text{ such that} \\ \langle A(u)y, z - y \rangle \geq \langle Bu + g, z - y \rangle \quad \text{for all} \quad z \in \tilde{K}, \end{cases}$$

where $\langle \, , \, \rangle$ denotes the duality pairing between V' and V.

Let us assume that $A(u): V \to V'$ is generated by a bilinear, continuous form $a_u : V \times V \to \mathbf{R}$:

$$\langle A(u)y, v \rangle = a_u(y, v), \qquad y, v \in V, \ u \in U_{\text{ad}} \tag{3.1}$$

satisfying with certain positive constants M and m (being independent of u) the following conditions:

$$|a_u(y,v)| \leq M\|y\|_V\|v\|_V \qquad\qquad \forall u \in U_{\text{ad}}, \ \forall y, v \in V \tag{3.2}$$
$$a_u(v,v) \geq m\|v\|_V^2 \qquad\qquad \forall v \in V, \ \forall u \in U_{\text{ad}} \tag{3.3}$$
$$a_u(y,v) = a_u(v,y) \qquad\qquad \forall y, v \in V, \ \forall u \in U_{\text{ad}} \tag{3.4}$$
$$u_n \to u \quad \text{in } U \implies A(u_n) \to A(u) \quad \text{in } L(V, V') \tag{3.5}$$

and eventually

$$u_n \rightharpoonup u \text{ in } U \text{ (weak convergence)} \implies A(u_n) \to A(u) \text{ in } L(V, V'), \tag{3.6}$$

where $L(V, V')$ denotes the space of linear, continuous mappings from V to V'.

By above assumptions the state problem $(S(u))$ has a unique solution $y = y(u)$ for any $u \in U_{\text{ad}}$. For the detailed proof see **Tiba** (1990), for example.

Let $J: U \times V \to \mathbf{R}$ be a convex, lower semicontinuous, proper functional. Then the following result is standard

Theorem 3.1. *Suppose that either a) or b) or c) is true:*

a) $U_{\text{ad}} \subset U$ *is compact and (3.5) holds;*

b) $K \subset V$ *is compact,* $U_{\text{ad}} \subset U$ *bounded, the mapping* $u \mapsto A(u)$ *is linear and continuous in* u *and (3.6) holds;*

c) $a(\cdot, \cdot)$ *is independent of* u, U_{ad} *is bounded and* $U \subset V'$ *is compact.*

Then **(P)** *has at least one optimal solution* (u^*, y^*) *if it has admissible pairs (i.e. if the set of* $u \in U_{ad}$ *such that* $y(u) \in K$ *is nonempty).*

Proof. The proof is presented in **Haslinger, Neittaanmäki and Tiba** (1987).

□

We give an example from **Mignot and Puel** (1984) where the system is described by a variational inequality and the control is on the right hand side. We shall see that the corresponding optimal control problem is nonsmooth.

Example 3.1. Consider the optimal control problem in **R**:

(P) $$\underset{u \in \mathbf{R}}{\text{minimize}} \ \left\{ J(u, y(u)) \equiv (y(u) - 1)^2 + u^2 \right\}$$

where $y = y(u) \in \widetilde{K}$ is the solution to the variational inequality

(S(u)) $$y(\varphi - y) \geq (-1 + u)(\varphi - v) \qquad \text{for all } \varphi \in \widetilde{K},$$

with $\widetilde{K} = [0, \infty)$. We find that

$$y(u) = (-1 + u)^+$$

is the solution to **(S(u))**. Consequently,

$$J(u, y(u)) = \begin{cases} 2u^2 - 4u + 4 & \text{for } u \geq 1 \\ u^2 + 1 & \text{for } u \leq 1. \end{cases} \tag{3.7}$$

Figures 3.1 and 3.2 show the solution to the state and the behaviour of the cost as a function of the control variable u.

This example indicates that control problems of systems governed by variational inequalities are, in general, nonsmooth. Moreover, we see that the problem is not only nonsmooth but also *nonconvex*.

3.3. Practical Model Problems

In order to motivate the study of the abstract control problem **(P)** we present several examples which fit into this setting. We shall use penalty technique (with exact penalty) for handling the state constraints. Our approach leads to an optimization problem with nonsmooth objective function. In **Haslinger and**

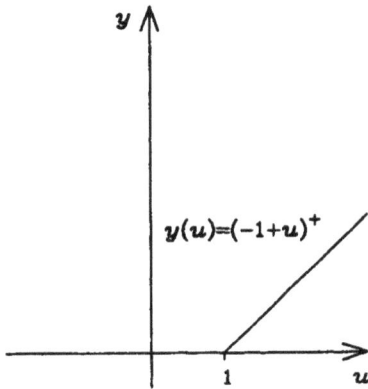

Figure 3.1.

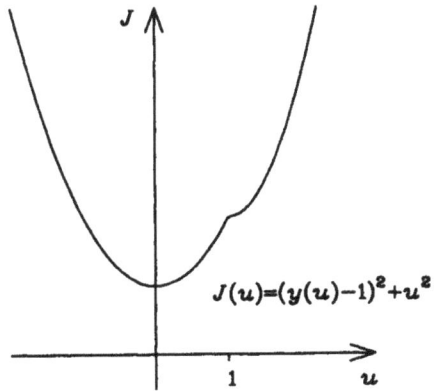

Figure 3.2.

Neittaanmäki (1988) similar problems were considered by applying the exterior penalty technique for the state constraints. In that case one can apply smooth optimization methods while the usage of exact penalty functions leads to non-smooth optimization. We shall use here both the exact and the exterior penalty techniques for handling the state constraints and compare the obtained numerical results.

3.3.1. Axially Loaded Rod with Stress Constraints

Consider an axially loaded rod (see Figure 3.3).

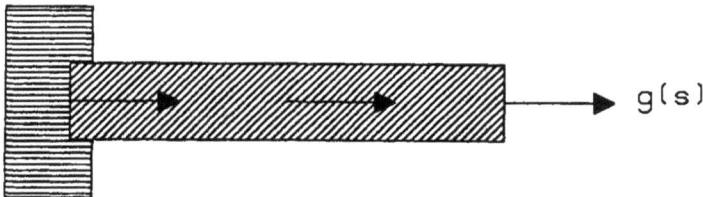

Figure 3.3.

Our goal is to minimize the weight of the rod. It is required that the stress does not exceed a given value $\sigma > 0$. Let u denote the radius of the rod, g the load and y the corresponding displacement at cartial line. We assume that the left endpoint is fixed (i.e. $y(0) = 0$) and the length of the rod is 1. By these assumptions the state system reads

$$(\mathbf{S_1(u)}) \qquad \begin{aligned} A(u)y &\equiv -(uy')' = g \qquad \text{in } (0,1), \\ y(0) &= 0, \quad y'(1) = 0. \end{aligned}$$

The control problem now takes the form

$$(\mathbf{P_1}) \qquad \underset{u \in U_{ad}}{\text{minimize}} \left\{ J(u) \equiv \int_0^1 u(s)\,ds \right\}$$

subject to $y = y(u)$ satisfying the state system $(\mathbf{S_1(u)})$ and the state constraint

$$|y'(s)| \leq \sigma \qquad \text{a.e. in } (0,1)$$

with a given constant $\sigma > 0$. The set of admissible controls is given by

$$U_{ad} = \{u \in W^{1,\infty}((0,1)) \mid 0 < \alpha \leq u(s) \leq \beta, \ |u'(s)| \leq \gamma \text{ a.e. in } (0,1)\}$$

with given positive constants α, β and γ. Referring to the abstract setting we have in this simple case

$$\begin{aligned} V &= \{v \in H^1((0,1)) \mid v(0) = 0\}, \quad U = C([0,1]), \\ K &= \{v \in V \mid |v'(s)| \leq \sigma \text{ a.e. in } (0,1)\}, \\ B &\equiv 0, \quad \varphi \equiv 0, \quad a_u(y,v) = \int_0^1 u(s)y'(s)v'(s)\,ds. \end{aligned}$$

For the numerical realization, let $0 = a_1 < a_2 < \cdots < a_{N(h)} = 1$ be an equidistant partition of $[0,1]$ with a step $h > 0$, i.e. $a_i = (i-1)h$, for $i = 1, \ldots, N(h)$, with $N(h) = (1+h)/h$. With this partition we associate the following finite dimensional spaces

$$\begin{aligned} U_{ad}^h &= \{u_h \in U_{ad} \mid u_h|_{[a_{i-1},a_i]} \in P_1, \ i = 2, \ldots, N(h)\}, \\ V_h &= \{v_h \in V \mid v_h|_{[a_{i-1},a_i]} \in P_1, \ i = 2, \ldots, N(h)\}, \end{aligned}$$

where P_1 denotes the space of polynomials of degree at most 1. The penalized (exact penalty) and discretized version of $(\mathbf{P}_1)$ reads as follows

$(\mathbf{P}_{\varepsilon h})$

$$\underset{u_h \in U_{ad}^h}{\text{minimize}} \left\{ J_{\varepsilon h}(u_h, y_h) \equiv \int_0^1 u_h(s)\, ds \right.$$

$$\left. + \frac{1}{\varepsilon} \int_0^1 \left[(y_h'(s) - \sigma)^+ + (-y_h'(s) - \sigma)^+ \right] ds \right\},$$

where $\varepsilon > 0$ and $y_h \in V_h$ is the solution of the state problem $(u_h \in U_{ad}^h)$:

$(\mathbf{S}_h(u_h))$ $\qquad\qquad a_{u_h}(y_h, v_h) = (g, v_h) \quad \forall v_h \in V_h.$

Here

$$a_{u_h}(y_h, v_h) = \int_0^1 u_h(s)y_h'(s)v_h'(s)\, ds, \quad y_h, v_h \in V_h$$

and

$$(g, v_h) = \int_0^1 g(s)v_h(s)\, ds.$$

In order to rewrite problem $(\mathbf{P}_{\varepsilon h})$ in terms of nodal values we note that

$$u_h = \sum_{i=1}^{N(h)} x_i \varphi_i$$

and

$$y_h = \sum_{i=1}^{N(h)} y_i \varphi_i$$

where

$$x_i = u_h(a_i) \quad \text{and} \quad y_i = y_h(a_i) \quad (y_1 = y_h(0) = 0)$$

for $i = 1, \ldots, N(h)$ denote the nodal values of u_h and y_h. Moreover, φ_i denotes the Courant's basis function, i.e. φ_i is piecewise linear and $\varphi_i(a_j) = \delta_{ij}$. Notice, that in this case we have $n = m = N(h)$ (the dimensions of the control and the state variables)[1].

---

[1] Because of notational convenience we do not eliminate the boundary values of the state problem, but we shall consider the fixed boundary values of the state as normal variables

We can now define analogously to the space U_{ad}^h:

$$\mathcal{U} = \{x \in \mathbf{R}^n \mid 0 < \alpha \le x_i \le \beta, \ i = 1, \ldots, n,$$
$$- \gamma h \le x_{i+1} - x_i \le \gamma h, \ i = 1, \ldots, n-1\}$$

and the cost functional $J_{\varepsilon h}(u_h, y_h)$ in terms of the nodal values as follows

$$J_\varepsilon(x, y(x)) = \frac{h}{2} \sum_{i=1}^{n-1}(x_i + x_{i+1}) + \frac{h}{\varepsilon} \sum_{i=1}^{n-1} \left[\left(\frac{y_{i+1} - y_i}{h} - \sigma \right)^+ + \left(-\frac{y_{i+1} - y_i}{h} - \sigma \right)^+ \right].$$

Moreover, $(\mathbf{S}_h(u_h))$ can be written:

$$(\mathcal{S}(x)) \qquad\qquad\qquad A(x)y = G,$$

where $A(x)$ is the stiffness matrix

$$A(x) = (A_{ij}(x))_{i,j=1}^n = (a_{u_h}(\varphi_i, \varphi_j))_{i,j=1}^n = \left(\int_0^1 u_h(s)\varphi_i'(s)\varphi_j'(s)\,ds \right)_{i,j=1}^n,$$

and G denotes the force vector

$$G = (G_j)_{j=1}^n = \left(\int_0^1 g(s)\varphi_j(s)\,ds \right)_{j=1}^n.$$

The matrix form of $(\mathbf{P}_{\varepsilon h})$ is now given by

$$(\mathcal{P}_\varepsilon) \qquad\qquad\qquad \underset{x \in \mathcal{U}}{\text{minimize}} \ J_\varepsilon(x, y(x))$$

subject to $y = y(x)$:

$$(\mathcal{S}(x)) \qquad\qquad\qquad A(x)y = G.$$

For the numerical realization we shall perform the sensitivity analysis by applying Theorem 2.2. We rewrite $J_\varepsilon(x, y(x))$ in the form

$$J_\varepsilon(x, y(x)) = h(x) + E(y(x))$$
$$\equiv C^{\mathrm{T}}x + \frac{1}{\varepsilon}\left[(\Phi(y(x)) - \Phi(-y(x)))^{\mathrm{T}} By(x) - h\,(\Phi(y(x)) + \Phi(-y(x)))^{\mathrm{T}} \boldsymbol{\sigma} \right].$$

Above $C = \frac{h}{2}(1, 2, 2, \ldots, 2, 1)^{\mathrm{T}} \in \mathbf{R}^n$ and $\Phi : \mathbf{R}^n \to \mathbf{R}^n$ with $\Phi_1 = 0$ and

$$\Phi_{i+1}(y) = \begin{cases} 1, & \text{if } \dfrac{y_{i+1} - y_i}{h} > \sigma \\ 0, & \text{otherwise.} \end{cases}$$

for $i = 1, \ldots, n - 1$, $(y_1 = 0)$,

$$B = \begin{pmatrix} 0 & 0 & 0 & \cdots & 0 & 0 \\ 0 & 1 & 0 & \cdots & 0 & 0 \\ 0 & -1 & 1 & \cdots & 0 & 0 \\ \vdots & \vdots & \vdots & \cdots & \vdots & \vdots \\ 0 & 0 & 0 & \cdots & -1 & 1 \end{pmatrix} \in \mathbf{R}^{n \times n}$$

and $\boldsymbol{\sigma} = (\sigma, \ldots, \sigma)^{\mathrm{T}} \in \mathbf{R}^n$. It is easy to verify that

$$\frac{1}{\varepsilon} \left(\Phi(y(x)) - \Phi(-y(x)) \right)^{\mathrm{T}} B \in \partial_y E(y(x))$$

and by Theorem 2.2

$$C + \frac{1}{\varepsilon} \nabla y(x)^{\mathrm{T}} \left[(\Phi(y(x)) - \Phi(-y(x)))^{\mathrm{T}} B \right] \in \partial J_\varepsilon(x, y(x)), \tag{3.8}$$

where

$$\nabla y(x) = \left(\frac{\partial}{\partial x_i} y_j(x) \right)_{i,j=1}^n \in \mathbf{R}^{n \times n}$$

denotes the Jacobian of $y(x)$.

Let $p(x) \in \mathbf{R}^n$ be the solution of the adjoint problem

$$(\mathcal{A}(x)) \qquad A(x)p(x) = (\Phi(y(x)) - \Phi(-y(x)))^{\mathrm{T}} B.$$

On the other hand, by differentiating the state problem $(\mathcal{S}(x))$ we obtain

$$\nabla A(x)[y(x)] + A(x)\nabla_x y(x) = 0, \tag{3.9}$$

where

$$\nabla A(x)[y(x)] = \left(\sum_{k=1}^n \frac{\partial}{\partial x_j} A_{ik}(x) y_k \right)_{i,j=1}^n \in \mathbf{R}^{n \times n}.$$

Consequently by (3.8) and (3.9) we get

$$C + \frac{1}{\varepsilon} \nabla_x y(x)^{\mathrm{T}} [(\Phi(y(x)) - \Phi(-y(x)))^{\mathrm{T}} B]$$
$$= C + \frac{1}{\varepsilon} \nabla_x y(x)^{\mathrm{T}} A(x)p(x) = C - \frac{1}{\varepsilon} \nabla A(x)[y(x)]^{\mathrm{T}} p(x)$$

i.e.

$$C - \frac{1}{\varepsilon}\nabla A(x)[y(x)]^{\mathrm{T}}p(x) \in \partial \mathcal{J}_\varepsilon(x, y(x)),$$

where $p(x)$ is a solution of the adjoint state $(\mathcal{A}(x))$.

Above we have applied the exact penalty technique and obtained the nonsmooth optimization problem $(\mathcal{P}_\varepsilon)$. In **Haslinger and Neittaanmäki** (1988) the problem $(\mathbf{P}_1)$ was solved by using the quadratic exterior penalty approach, which leads to the smooth optimization problem

$$(\mathcal{P}_\varepsilon^{ext}) \qquad \begin{aligned} \underset{x \in \mathcal{U}}{\text{minimize}} &\left\{ \mathcal{J}_\varepsilon^{ext}(x, y(x)) \equiv \frac{h}{2}\sum_{i=1}^{n-1}(x_i + x_{i+1}) \right. \\ &\left. + \frac{h}{2\varepsilon}\sum_{i=1}^{n-1}\left[\left[\left(\frac{y_{i+1}-y_i}{h} - \sigma \right)^+ \right]^2 + \left[\left(-\frac{y_{i+1}-y_i}{h} - \sigma \right)^+ \right]^2 \right] \right\} \end{aligned}$$

subject to $y = y(x)$ solving the state problem $(\mathcal{S}(x))$.

The sensitivity analysis for the problem $(\mathcal{P}_\varepsilon^{ext})$ can be performed in a straightforward manner by applying the adjoint state technique (see **Haslinger and Neittaanmäki** (1988)).

The nonsmooth problem $(\mathcal{P}_\varepsilon)$ has been solved by the code PB and for the problem $(\mathcal{P}_\varepsilon^{ext})$ we have applied the sequential quadratic programming (SQP) method (subroutine E04VCF from NAG-library). The state problem $(\mathcal{S}(x))$ and the adjoint state problem $(\mathcal{A}(x))$ were solved by Cholesky factorization.

Let the rod be loaded by a uniform load $g(s) \equiv 1$. We used 16 elements in discretization, i.e. $h = 1/16$ and $n = 17$. Let the parameters in the definition of $\mathcal{U}$ be given by $\alpha = 0.025$, $\beta = 0.1$ and $\gamma = 0.1$ and let the initial guess be $x_i^1 = \beta$ for $i = 1, \ldots, 17$, which is feasible with respect to the state constraints. Figure 3.4 shows the lengthwise section of the rod with these parameters. Furthermore, let $\sigma = 10$, then the maximum stress occurs at $s = 0$ and equals to σ. The initial 'volume' (value of the cost without penalty) of the rod is 0.1. The test runs have been performed in HP 9000/425 work station with double precision relative accuracy $\approx 10^{-16}$. In order to obtain feasible solutions the penalty parameters were chosen to be $\varepsilon = 10^{-3}$ for both exact penalty and exterior penalty.

Figure 3.5 contains the decrease of the volume and the cost functions $\mathcal{J}_\varepsilon$ and $\mathcal{J}_\varepsilon^{ext}$ versus the serious steps obtained by PB (see Algorithm II 3.2.2) and E04VCF codes, respectively. In Figure 3.6 (a) and (b) we see the decrease of the cost functions and the volumes obtained by both methods. Notice that if these cost $=$ volume in Figure 3.6, then the penalty term is zero, i.e. the solution is feasible

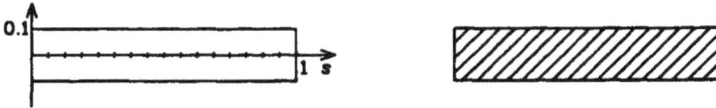

Figure 3.4.

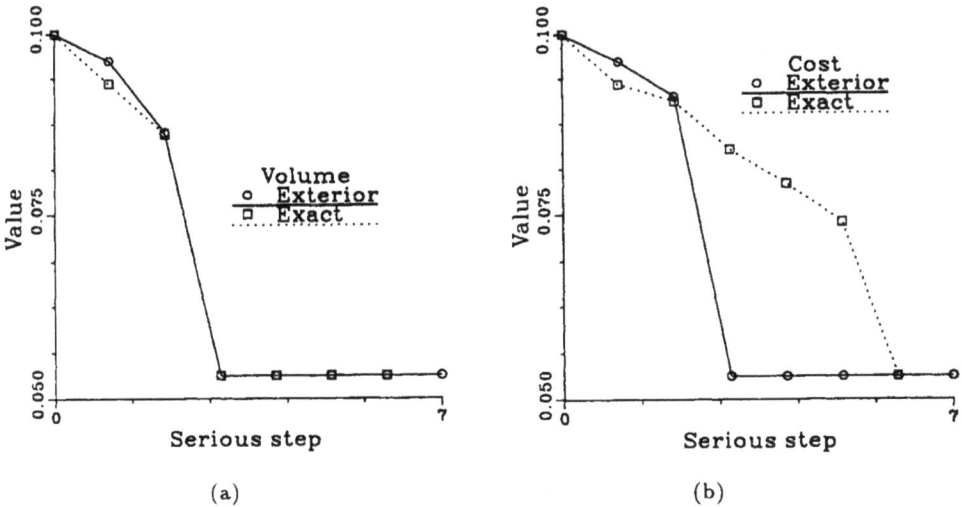

(a)   (b)

Figure 3.5.

with respect to state constraints. By both methods we obtained the same shape for the rod and it is illustrated in Figure 3.7.

Our results are summarized in Table 3.1. The following abbreviations will be used: **Initial** = initial value of the cost function, **Final** = final value of the cost function, **it** = number of iterations and **nf** = number of function and subgradient (gradient) evaluations.

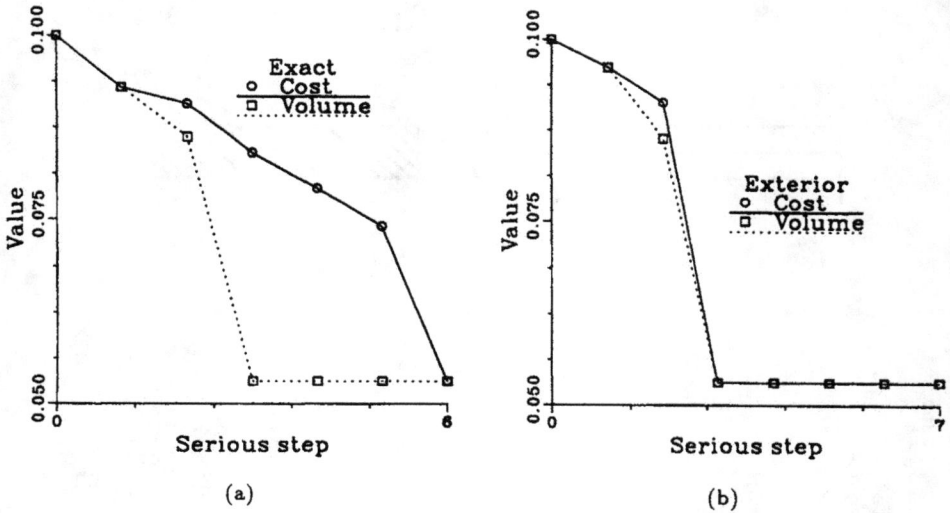

Figure 3.6.

Figure 3.7.

Table 3.1.

Cost	Initial	Final	it	nf	Algorithm
J_e	0.100000	0.053125	6	17	PB
J_e^{ext}	0.100000	0.053125	7	36	E04VCF

In both cases the state constraints are satisfied and the lower bound $\alpha \leq x_i$ becomes active for $i = 12, \ldots, 17$. Figure 3.8 shows the corresponding result obtained by exact penalty method when we have made the lower bound to be $\alpha = 0.0125$. PB needed 20 iterations (9 serious steps) and 21 function evaluations and the final cost was 0.05078.

Figure 3.8.

3.3.2. *Clamped Beam with Displacement Constraints*

Consider the model of a clamped beam, subject to vertical load g (see Figure 3.9).

Let u be the thickness of the beam, $b > 0$ a constant depending on the shape of the cross section of the beam and on Young's modulus of the material used. It is further required that the deflection y of the beam remains within certain limits, say $|y| \leq r$.

$$g(s)$$

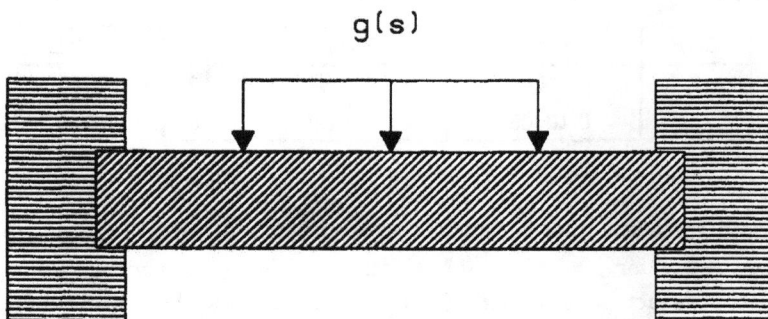

Figure 3.9.

In this case the state problem reads

$(\mathbf{S_2(u)})$
$$A(u)y \equiv (bu^3 y'')'' = g \qquad \text{in } (0,1)$$
$$y(0) = y'(0) = y(1) = y'(1) = 0.$$

The control problem now takes the form

$(\mathbf{P_2})$
$$\underset{u \in U_{ad}}{\text{minimize}} \left\{ J(u) \equiv \int_0^1 u(s)\,ds \right\}$$

subject to $y = y(u)$ satisfying the state system $(\mathbf{S_2(u)})$ and the state constraint

$$|y(s)| \leq r \qquad \text{a.e. in } [0,1]$$

with a given constant $r > 0$. Above the set of admissible controls is given by

$$U_{ad} = \{ u \in W^{1,\infty}((0,1)) \mid 0 < \alpha \leq u(s) \leq \beta,\ |u'(s)| \leq \gamma \text{ a.e. in } (0,1) \}$$

with given positive constants α, β and γ. In this case

$$V = H_0^2((0,1)), \quad U = C([0,1]),$$
$$K = \{ v \in V \mid |v(s)| \leq r \text{ in } [0,1] \},$$
$$B \equiv 0, \quad \varphi \equiv 0, \quad a_u(y,v) = \int_0^1 bu^3(s)y''(s)v''(s)\,ds.$$

Again let $0 = a_1 < a_2 < \cdots < a_{N(h)} = 1$ be an equidistant partition of $[0, 1]$, $h = a_{i+1} - a_i$. With this partition we associate sets $(i = 2, \ldots, N(h) = (1+h)/h)$

$$U_{ad}^h = \{u_h \in C([0,1]) \mid u_h|_{[a_{i-1}, a_i]} \in P_1\} \cap U_{ad},$$
$$V_h = \{v_h \in C^1([0,1]) \mid v_h|_{[a_{i-1}, a_i]} \in P_3, \ v_h(0) = v_h'(0) = v_h(1) = v_h'(1) = 0\}.$$

Functions from V_h are uniquely determined by their values and the values of their first derivatives at nodes a_i, $i = 2, \ldots, N(h) - 1$.

The penalized (exact penalty) and discretized version of $(\mathbf{P}_2)$ now reads as follows

$(\mathbf{P}_{\varepsilon h})$
$$\operatorname*{minimize}_{u_h \in U_{ad}^h} \left\{ J_{\varepsilon h}(u_h, y_h) \equiv \int_0^1 u_h(s)\, ds \right.$$
$$\left. + \frac{1}{\varepsilon} \int_0^1 \left[(y_h(s) - r)^+ + (-y_h(s) - r)^+\right] ds \right\},$$

where $\varepsilon > 0$ and $y_h \in V_h$ is the solution of the state problem

$(\mathbf{S}(u_h))$ $\qquad\qquad a_{u_h}(y_h, v_h) = (g, v_h) \quad$ for all $v_h \in V_h$

and

$$a_{u_h}(y_h, v_h) = \int_0^1 b u_h^3(s) y_h''(s) v_h''(s)\, ds, \qquad y_h, v_h \in V_h,$$

$$(g, v_h) = \int_0^1 g(s) v_h(s)\, ds.$$

The matrix form of $(\mathbf{P}_{\varepsilon h})$ now reads $(n = N(h))$

$(\mathcal{P}_\varepsilon)$
$$\operatorname*{minimize}_{x \in \mathcal{U}} \left\{ \mathcal{J}_\varepsilon(x, y(x)) \equiv \frac{h}{2} \sum_{i=1}^{n-1} (x_i + x_{i+1}) \right.$$
$$\left. + \frac{1}{\varepsilon} \sum_{i=2}^{n-1} \omega_i \left[(y_{2i-1} - r)^+ + (-y_{2i-1} - r)^+\right] \right\},$$

subject to $y = y(x)$

$(\mathcal{S}(x))$ $\qquad\qquad A(x)y = G,$

where $x_i = u_h(a_i)$ for $i = 1, \ldots, n$, $\mathcal{U}$ is the same as in Subsection 3.3.1 and $\omega_i \in \mathbf{R}$ are the weights of the formula, used for the numerical computation of the nonlinear penalty term. Moreover, $y = (y_h(a_1), y_h'(a_1), \ldots, y_h(a_n), y_h'(a_n))^T \in \mathbf{R}^{2n}$ is the vector, containing degrees of freedom of the solution $y_h \in V_h$ at nodes $a_2, \ldots, a_{n-1}$, i.e. $y_{2i-1} = y_h(a_i)$, $y_{2i} = y_h'(a_i)$, and the fixed boundary values $y_1 = y_2 = y_{2n-1} = y_{2n} = 0$. Thus, in this example $m = 2n$. Moreover, $A(x) \in \mathbf{R}^{m \times m}$ is the stiffness matrix and $G \in \mathbf{R}^m$ is the force vector of the discrete problem $(\mathcal{S}(u_h))$.

For the numerical realization we shall again perform the sensitivity analysis following similar lines as in Subsection 3.3.1. We write $\mathcal{J}_\varepsilon(x, y(x))$ in the form

$$\mathcal{J}_\varepsilon(x, y(x)) = h(x) + E(y(x))$$

$$\equiv C^T x + \frac{1}{\varepsilon} \left[(\Phi(y(x)) - \Phi(-y(x))) B y(x) - (\Phi(y(x)) + \Phi(-y(x))) \mathbf{r} \right],$$

where $\Phi : \mathbf{R}^m \to \mathbf{R}^m$ with

$$\Phi_{2i}(y) = 0, \quad \text{for} \quad i = 1, \ldots, n$$

$$\Phi_{2i-1}(y) = \begin{cases} 1, & \text{if } y_{2i-1} > r, \ i = 1, \ldots, n \\ 0, & \text{otherwise,} \end{cases}$$

and

$$C = \frac{h}{2}(1, 2, \ldots, 2, 1)^T \in \mathbf{R}^n,$$

$$B = \operatorname{diag}(\omega_1, 0, \omega_2, \ldots, \omega_n, 0) \in \mathbf{R}^{m \times m},$$

$$\mathbf{r} = (r, r, \ldots, r)^T \in \mathbf{R}^m.$$

As

$$\frac{1}{\varepsilon}(\Phi(y(x)) - \Phi(-y(x)))^T B \in \partial_y E(y(x))$$

we obtain by Theorem 2.2

$$C + \frac{1}{\varepsilon} \nabla y(x)^T \left[(\Phi(y(x)) - \Phi(-y(x)))^T B \right] \in \partial \mathcal{J}_\varepsilon(x, y(x)). \tag{3.10}$$

Let $p(x) \in \mathbf{R}^m$ be the solution of the adjoint state

$$(\mathcal{A}(x)) \qquad\qquad A(x) p(x) = (\Phi(y(x)) - \Phi(-y(x)))^T B.$$

Differentiating the state problem $(\mathcal{S}(x))$ we obtain with similar notations as in Subsection 3.3.1

$$\nabla A(x)[y(x)] + A(x) \nabla_x y(x) = 0. \tag{3.11}$$

Now (3.10) and (3.11) yield

$$C - \frac{1}{\varepsilon}\nabla A(x)[y(x)]^{\mathrm{T}} p(x) \in \partial J_\varepsilon(x, y(x)), \tag{3.12}$$

where $p(x)$ solves the adjoint state $(\mathcal{A}(x))$.

We have solved numerically the problem $(\mathbf{P_2})$ by applying the exact penalty method as well as the quadratic exterior penalty method. In the latter case we have solved the smooth optimization problem

$(\mathcal{P}_\varepsilon^{ext})$
$$\underset{x \in \mathcal{U}}{\text{minimize}} \left\{ J_\varepsilon^{ext}(x, y(x)) \equiv \frac{h}{2} \sum_{i=1}^{n-1} (x_i + x_{i+1}) \right.$$
$$\left. + \frac{h}{2\varepsilon} \sum_{i=1}^{n-1} \omega_i \left[\left[(y_{2i-1} - r)^+ \right]^2 + \left[(-y_{2i-1} - r)^+ \right]^2 \right] \right\}$$

subject to $y = y(x)$ solving the state problem $(\mathcal{S}(x))$.

The nonsmooth problem $(\mathcal{P}_\varepsilon)$ has been solved by the code PB and for the problem $(\mathcal{P}_\varepsilon^{ext})$ we have applied the sequential quadratic programming (SQP) method (subroutine E04VCF from NAG-library). The state problem $(\mathcal{S}(x))$ and the adjoint state problem $(\mathcal{A}(x))$ were solved by Cholesky factorization.

Let $g(s) \equiv -1$, $n = 31$ (i.e. $h = 1/30$ in FE-discretization), $\alpha = \frac{1}{10}$, $\beta = 1$, $\gamma = 2$ and $r = \frac{1}{100}$. Let the initial guess be $x_i^1 = \beta$ for $i = 1, \ldots, 31$, which does not violate the state constraints (see Figure 3.10). The test runs have been performed in HP 9000/425 work station with double precision relative accuracy $\approx 10^{-16}$. In order to obtain feasible solutions (with respect to state constraints) the penalty parameters were chosen to be $\varepsilon = 10^{-4}$ and $\varepsilon = 10^{-8}$ for exact penalty and for exterior penalty, respectively.

Figure 3.11 contains the decrease of the volume and the cost functions J_ε and J_ε^{ext} versus the serious steps obtained by PB and E04VCF codes, respectively. In Figure 3.12 (a) and (b) we see the decrease of the cost functions and the volumes obtained by both methods. By both methods we obtained the same final shape for the beam which is illustrated in Figure 3.13 (a). In Figure 3.13 (b) we see the deflection of the beam.

Our results are summarized in Table 3.2.

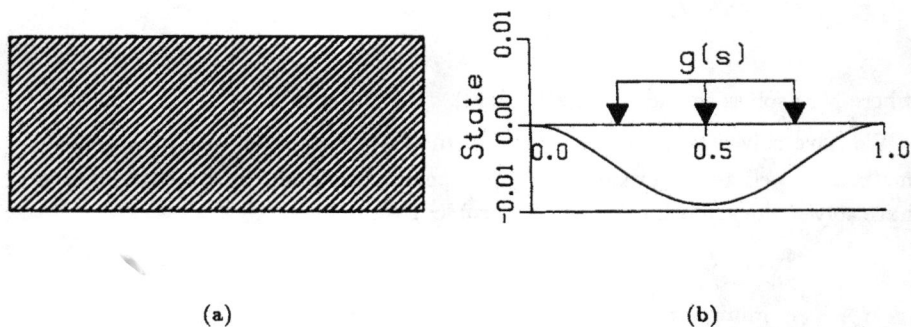

(a)   (b)

Figure 3.10.

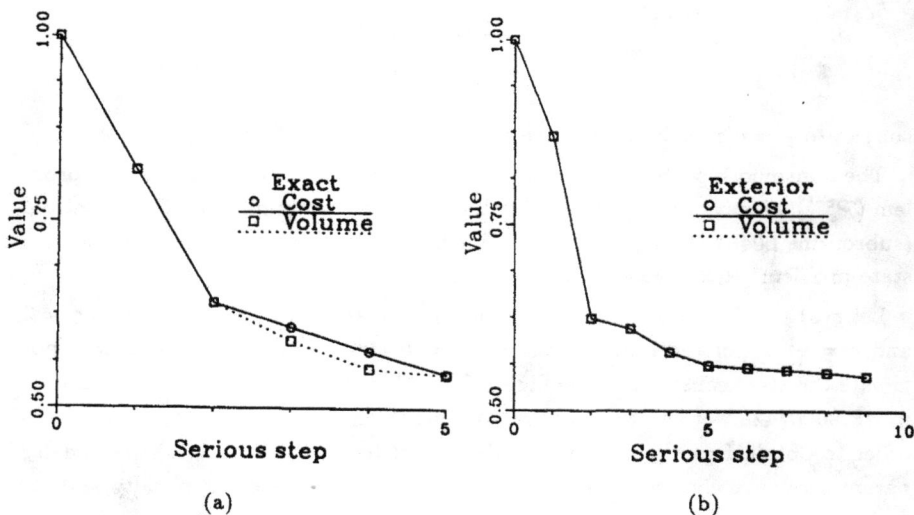

(a)   (b)

Figure 3.11.

Table 3.2.

Cost	Initial	Final	it	nf	Algorithm
$\mathcal{J}_\varepsilon$	1.000000	0.546736	16	17	PB
$\mathcal{J}_\varepsilon^{ext}$	1.000000	0.546740	9	34	E04VCF

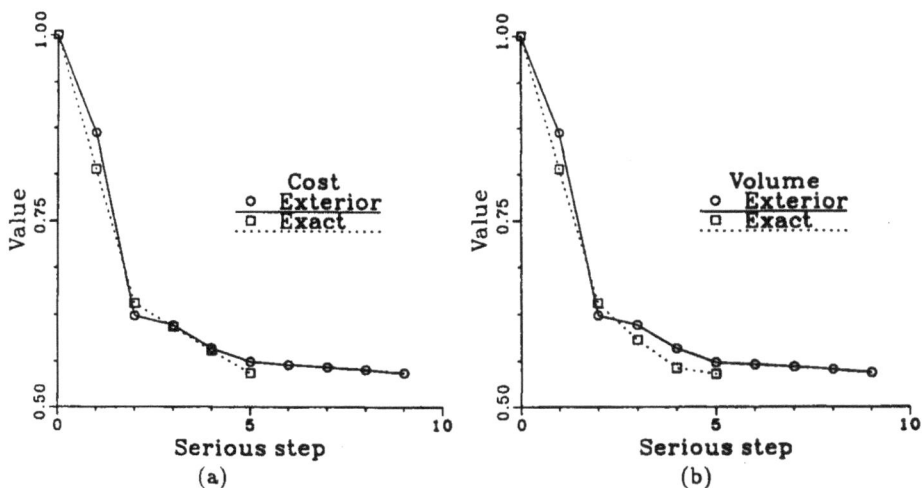

Figure 3.12.

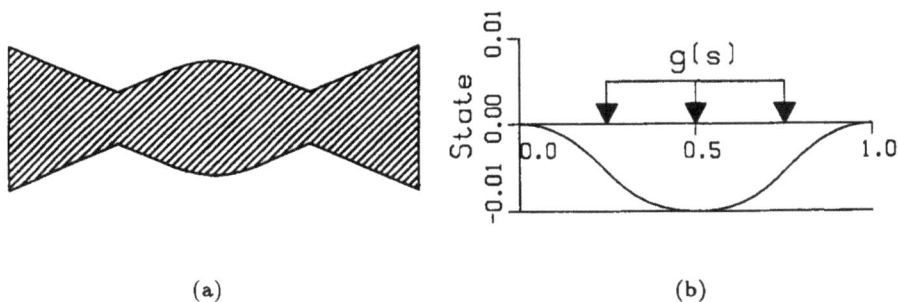

(a)   (b)

Figure 3.13.

Remark 3.2. Instead of the deflection constraint $|y(s)| \leq r$ in $[0, 1]$ one can consider the case where the bending stress in the beam may not exceed a given value δ, i.e.

$$K = \{y \in V \mid \|bu^3 y''\|_{L^\infty((0,1))} < \delta\}.$$

The treatment of this problem is analogous to Subsection 3.3.2.

3.3.3. Clamped Beam with Obstacle

Let us consider a clamped beam the deflection of which is limited from below by a rigid obstacle described by the graph of a function q. The control variable expresses the physical meaning of the load of the beam. The aim is to find a load density u in such a way that the area between the beam and the obstacle will be minimized.

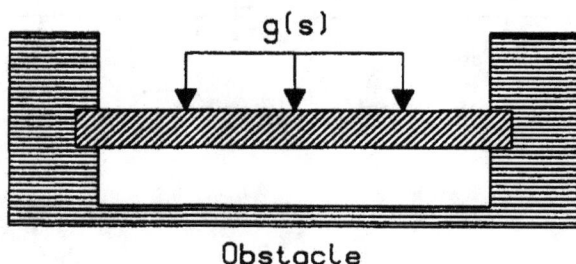

Figure 3.14.

Here

$$V = H_0^2((0,1)), \quad K = V, \quad U = L^2((0,1))$$

$$U_{ad} = \{u \in L^\infty((0,1)) \mid \alpha \le u(s) \le \beta, \text{ a.e. in } (0,1), \int_0^1 u(s)\,ds = M\}$$

with given positive constants α, β and M,

$\varphi: V \to (-\infty, +\infty]$ is given by
$$\varphi(y) = \begin{cases} 0 & y \ge q \text{ a.e. in } (0,1) \\ +\infty & \text{otherwise} \end{cases}$$
$A(u)y = y^{(iv)} \ (= y'''')$ (independent of u)
$B: U \to H^{-2}((0,1))$ is the natural embedding operator.

In this case the state problem can be formulated as a variational inequality to find $y(u) \in \tilde{K}$ such that

$(\mathbf{S_3}(u))$ $\qquad a(y(u), v - y(u)) \geq (g + u, v - y(u)) \qquad$ for all $\quad v \in \tilde{K}$,

where

$$\tilde{K} = \{v \in H_0^2((0,1)) \mid v \geq q \text{ a.e. in } (0,1)\}$$

and

$$a(y, v) = \int_0^1 y''(s) v''(s) \, ds$$

$$(g, v) = \int_0^1 g(s) v(s) \, ds.$$

The optimization problem reads

$(\mathbf{P_3})$ $\qquad \underset{u \in U_{ad}}{\text{minimize}} \left\{ J(y(u)) \equiv \int_0^1 (y(s) - q(s))^2 \, ds \right\},$

where $y \in \tilde{K}$ solves $(\mathbf{S_3}(u))$.

For the numerical realization let in this case $0 = a_0 < a_1 < \cdots < a_{N(h)} = 1$ to be an equidistant partition of $[0,1]$ and define

$$U_{ad}^h = \{u_h \in L^2((0,1)) \mid u_h|_{[a_{i-1}, a_i]} \in P_0, \ i = 1, \ldots, N(h)\} \cap U_{ad},$$
$$V_h = \{v_h \in C^1([0,1]) \mid v_h|_{[a_{i-1}, a_i]} \in P_3, \ v_h(0) = v_h'(0) = v_h(1) = v_h'(1) = 0\},$$
$$\tilde{K}_h = \{v_h \in V_h \mid v_h(a_i) \geq q(a_i), \ i = 1, \ldots, N(h)\}.$$

The discrete form of this optimal control problem reads

$(\mathbf{P_h})$ $\qquad \underset{u_h \in U_{ad}^h}{\text{minimize}} \left\{ J_h(y_h(u_h)) \equiv \int_0^1 (y_h(s) - q(s))^2 \, ds \right\},$

where $y_h \in \tilde{K}_h$ is the solution of

$(\mathbf{S_h}(u_h))$ $\qquad a(y_h, z_h - y_h) \geq (g + u_h, z_h - y_h) \qquad$ for all $\quad z_h \in \tilde{K}_h$.

Using the algebraic expression of J_h we obtain

$$\mathcal{J}(y(x)) = \omega_1 q_1^2 + \sum_{i=2}^{n-1} \omega_i (y_{2i-1} - q_i)^2 + \omega_n q_n^2 \qquad (y_1 = y_{2n} = 0),$$

where $n = N(h)$, $x_i = u_h(a_i)$ for $i = 1, \ldots, n$, ω_i are again the weights of a quadrature formula used for the calculation of J_h, $y = y(x) \in \mathbf{R}^m$ has exactly the same meaning as in the last example, and $q_i = q(a_i)$ for $i = 1, \ldots, n$. Thus also in this case $m = 2n$.

The approximate problem $(\mathbf{P}_h)$ in matrix form reads

$$(\mathcal{P}) \qquad\qquad \underset{x \in \mathcal{U}}{\text{minimize }} \mathcal{J}(y(x)),$$

where $y \equiv y(x)$ is the solution of the quadratic programming problem

$$(\mathcal{S}(x)) \qquad\qquad \underset{y \in \mathcal{K}}{\text{minimize }} \left\{ \tfrac{1}{2} y^\mathrm{T} A y - G(x)^\mathrm{T} y \right\}.$$

Here $A \in \mathbf{R}^{m \times m}$ is the stiffness matrix and $G(x) \in \mathbf{R}^m$ is the force vector of the discrete problem $(\mathbf{S}_h(u_h))$. Moreover,

$$\mathcal{U} = \{x \in \mathbf{R}^n \mid \alpha \leq x_i \leq \beta, \ i = 1, \ldots, n; \ \sum_{i=1}^{n} h x_i = M \},$$

$$\mathcal{K} = \{y \in \mathbf{R}^m \mid y_{2i-1} \geq q_i, \ i = 2, \ldots, n-1, \ y_1 = y_2 = y_{m-1} = y_m = 0 \}.$$

Let us note that in Problem $(\mathcal{S}(x))$ only the right hand side G depends on the control variable x.

In order to perform the sensitivity analysis we rewrite the cost functional in matrix form

$$\mathcal{J}(y(x)) = (y(x) - Q)^\mathrm{T} B (y(x) - Q), \tag{3.13}$$

where

$$B = \mathrm{diag}(\omega_1, 0, \omega_2, 0, \ldots, \omega_{n-1}, 0, \omega_n, 0) \in \mathbf{R}^{m \times m}$$

$$Q = (q_1, 0, q_2, 0, \ldots, q_{n-1}, 0, q_n, 0) \in \mathbf{R}^m.$$

According to Theorem 2.1

$$\xi_y(x)^\mathrm{T} \nabla_y \mathcal{J}(y(x)) = 2 \xi_y(x)^\mathrm{T} B (y(x) - Q) \in \partial \mathcal{J}(y(x)), \tag{3.14}$$

where $\xi_y(x) \in \partial y(x)$.

We shall now analyse how to apply an adjoint state technique for eliminating $\xi_y(x)$ in (3.14). Let $I = I_y \cup I_{y'} \cup I_\Gamma$ be the set of indices where

$$I_y = \{3, 5, \ldots, m-3\}$$
$$I_{y'} = \{4, 6, \ldots, m-2\}$$
$$I_\Gamma = \{1, 2, m-1, m\}.$$

Introducing a vector of non-negative Lagrange multipliers

$$\lambda(x) = (\lambda_1, \ldots, \lambda_m)$$

we can express $(S(x))$ in an equivalent way using the classical Karush–Kuhn–Tucker optimality conditions:

$$
\begin{aligned}
Ay(x) &= G(x) + \lambda(x) \\
(y_i(x) - q_{2i-1})\lambda_i(x) &= 0 \qquad \text{for all} \quad i \in I_y \\
\lambda_i(x) &= 0 \qquad \text{for all} \quad i \in I_{y'} \cup I_\Gamma.
\end{aligned}
\tag{3.15}
$$

We divide I_y into three parts $I_a(x)$, $I_n(x)$ and $I_s(x)$ given by

$$I_a(x) = \{i \in I_y \mid y_i(x) = q_{2i-1} \text{ and } \lambda_i(x) > 0\} \tag{3.16}$$
$$I_n(x) = \{i \in I_y \mid y_i(x) > q_{2i-1} \text{ and } \lambda_i(x) = 0\} \tag{3.17}$$
$$I_s(x) = \{i \in I_y \mid y_i(x) = q_{2i-1} \text{ and } \lambda_i(x) = 0\}. \tag{3.18}$$

We call $I_a(x)$, $I_s(x)$ and $I_n(x)$ the set of active, semiactive and nonactive constraints at x. The next result follows from Appendix II of **Haslinger and Neittaanmäki** (1988).

Theorem 3.2. *Let $y(x) \in \mathcal{K}$ be the solution of the state problem $(S(x))$. Then the directional derivative $y'(x; d) = (y_1'(x; d), \ldots, y_m'(x; d))^T$ at $x \in \mathcal{U}$ in the direction $d \in \mathbf{R}^n$ exists and can be found as the unique solution of the quadratic programming problem*

$$\underset{z \in \tilde{\mathcal{K}}(x)}{\text{minimize}} \left\{ \tfrac{1}{2} z^T A z - z^T \nabla G(x) d \right\}, \tag{3.19}$$

where

$$
\tilde{\mathcal{K}}(x) = \{z \in \mathbf{R}^m \mid z_i = 0 \quad \text{for all} \quad i \in I_a(x) \\
z_i \geq 0 \quad \text{for all} \quad i \in I_s(x)\}.
$$

If $I_s(x) \neq \emptyset$ then $y'(x; d)$ is nonlinear in d and therefore y is only directionally differentiable at x. On the other hand, if for $x \in \mathcal{U}$ the set $I_s(x) = \emptyset$, $y'(x; d)$ is linear in d and $x \mapsto y(x)$ is differentiable at x and therefore $x \mapsto \mathcal{J}(y(x))$ is differentiable at x, too.

In order to eliminate the term $\xi_y(x)$ in (3.14), we introduce the adjoint state $p(x) \in \mathbf{R}^m$ (if exists) to Problem $(\mathcal{P})$ as a solution of the quadratic programming problem

$$(\mathcal{A}(x)) \qquad \underset{p \in \tilde{\mathcal{K}}(x)}{\text{minimize}} \; \left\{ \tfrac{1}{2} p^{\mathrm{T}} A p - p^{\mathrm{T}} B(y(x) - Q) \right\}.$$

Now we can apply the Karush–Kuhn–Tucker optimality conditions to $(\mathcal{A}(x))$ and we obtain a vector of Lagrange multipliers $\mu(x) \in \mathbf{R}^m$ such that

$$\begin{aligned}
Ap(x) &= B(y(x) - Q) + \mu(x) \\
p_i(x) &= 0 \qquad \text{for all} \quad i \in I_a(x) \\
\mu_i(x) &= 0 \qquad \text{for all} \quad i \in I_n(x) \\
\mu_i(x) \geq 0, p_i(x) \geq 0, p_i(x)\mu_i(x) &= 0 \qquad \text{for all} \quad i \in I_s(x).
\end{aligned} \tag{3.20}$$

By using (3.14) and (3.20) we obtain

$$2\xi_y(x)^{\mathrm{T}}(Ap(x) - \mu(x)) = 2(A\xi_y(x))^{\mathrm{T}} p(x) - \xi_y(x)^{\mathrm{T}}\mu(x) \in \partial\mathcal{J}(y(x)). \tag{3.21}$$

Due to (3.15) we have

$$\partial(Ay(x)) = \partial G(x) + \partial\lambda(x),$$

which implies that

$$A\xi_y(x) = \nabla G(x) + \xi_\lambda(x)$$

for some $\xi_\lambda(x) \in \partial\lambda(x)$. Consequently,

$$2(\nabla G(x) + \xi_\lambda(x))^{\mathrm{T}} p(x) - \xi_y(x)^{\mathrm{T}}\mu(x) \in \partial\mathcal{J}(y(x)). \tag{3.22}$$

If $i \in I_a(x)$, then $p_i(x) = 0$ by (3.20) and by continuity of y there exists $t > 0$ such that $y_i(x + td) = q_{2i-1}$ for all $d \in \mathbf{R}^n$. Thus $y_i'(x; d) = 0$. On the other hand if $i \in I_n(x)$ then $\mu_i(x) = 0$ by (3.20) and by continuity of λ there exists $t > 0$ such that $\lambda_i(x + td) = 0$ for all $d \in \mathbf{R}^n$. Thus $\lambda_i'(x; d) = 0$. We can conclude that, if $i \in I_a(x) \cup I_n(x)$, then $\xi_\lambda(x)^{\mathrm{T}} p(x) = 0$ and $\xi_y(x)^{\mathrm{T}}\mu(x) = 0$. Thus, if $I_s(x) = \emptyset$, then we have

$$2\nabla G(x)^{\mathrm{T}} p(x) \in \partial\mathcal{J}(y(x)). \tag{3.23}$$

Under some technical assumptions it is proved in **Outrata** (1990) that (3.23) also holds in the case $I_s(x) \neq \emptyset$.

In this case the nonsmooth problem $(\mathcal{P}_\varepsilon)$ has not been solved only by the code PB, but also by the "smooth code" E04VCF from NAG-library, since according to **Haslinger and Neittaanmäki** (1988) the application of the regularization technique did not change the behaviour of the optimization routine. The state problem $(\mathcal{S}(x))$ and the adjoint state problem $(\mathcal{A}(x))$ were solved by Powell's quadratic solver ZQPCVX.

Let $\alpha = 0$, $\beta = 10$, $M = 5$ and the initial guess be $x_i^1 = M$ for $i = 1, \ldots, n$. Further, let $q \equiv -0.001$ describe the obstacle. In the first example we have discretized the problem by using 40 elements, i.e. $h = 1/40$ and $n = 40$. The test runs have been performed in VAX 4000 computer with double precision relative accuracy $\approx 10^{-17}$.

The initial situtation is presented in Figure 3.15.

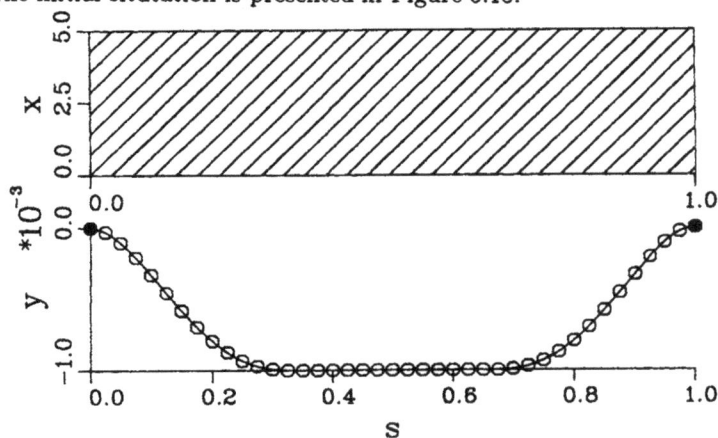

Figure 3.15.

Figure 3.16 contains the decrease of the cost functions versus the serious steps obtained by PB and E04VCF codes, respectively. In Figures 3.17 and 3.18 we see the obtained optimal controls and corresponding solutions of the state problem by PB and E04VCF, respectively. In this case the final solutions are slightly different and the value of the cost function obtained by PB is clearly lower. Note also that the optimal solution obtained by PB deviates from those obtained by BT code in **Schramm and Zowe** (1990), where the optimal cost value was $1.668395 \cdot 10^{-2}$ as in our case it was $1.668068 \cdot 10^{-2}$.

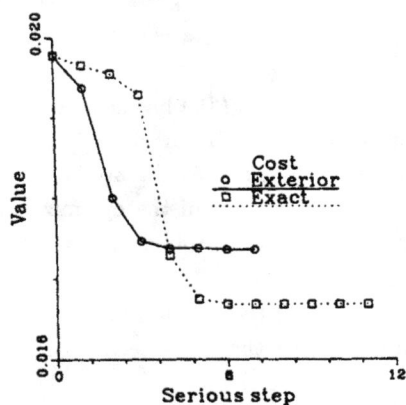

Figure 3.16.

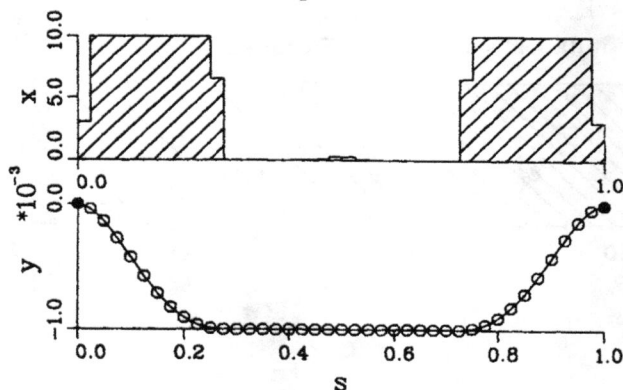

Figure 3.17. Optimal solution of PB

Our results are summarized in Table 3.3.

Table 3.3.

Cost	Initial	Final	it	nf	Algorithm
$\mathcal{J}_\varepsilon$	$1.977989 \cdot 10^{-2}$	$1.668068 \cdot 10^{-2}$	43	44	PB
$\mathcal{J}_\varepsilon$	$1.977989 \cdot 10^{-2}$	$1.735581 \cdot 10^{-2}$	6	16	E04VCF

Next we shall analyse the dependence of the optimal control on the discretization parameter. All the other parameters are the same as before, only the dimension

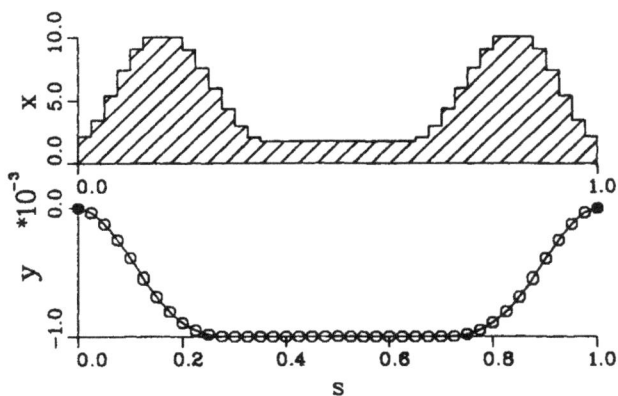

Figure 3.18. Optimal solution of E04VCF

changes; we have solved the problem by PB with $n = 10, 20, 50, 80$. The results are illustrated in Figures 3.19–3.22.

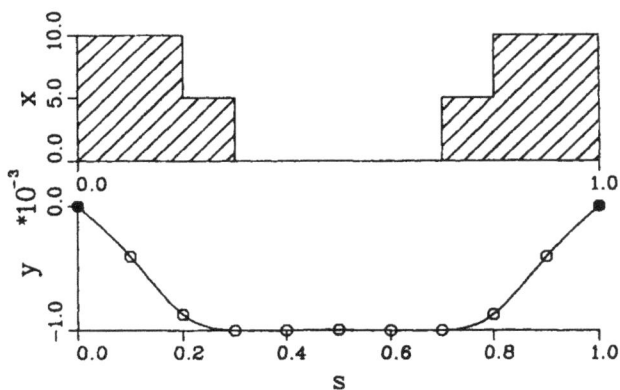

Figure 3.19. $n = 10$

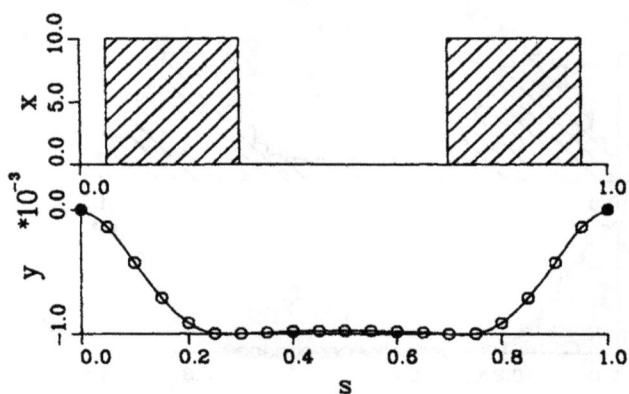

Figure 3.20. $n = 20$

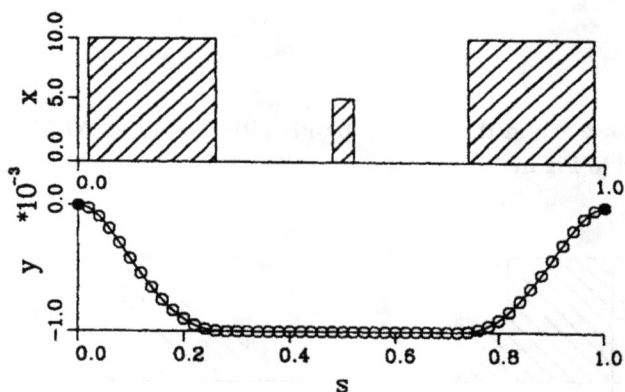

Figure 3.21. $n = 50$

Our results are summarized in Table 3.4.

Table 3.4.

Dim	Initial	Final	it	nf	Algorithm
$n = 10$	$1.990488 \cdot 10^{-2}$	$1.744498 \cdot 10^{-2}$	5	6	**PB**
$n = 20$	$1.978328 \cdot 10^{-2}$	$1.675682 \cdot 10^{-2}$	6	7	**PB**
$n = 50$	$1.980390 \cdot 10^{-2}$	$1.668244 \cdot 10^{-2}$	43	44	**PB**
$n = 80$	$1.979067 \cdot 10^{-2}$	$1.668291 \cdot 10^{-2}$	7	8	**PB**

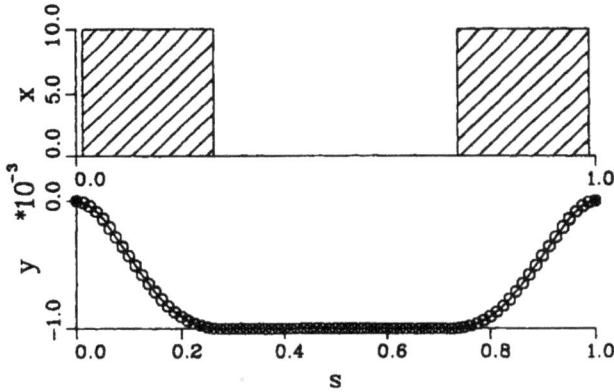

Figure 3.22. $n = 80$

Remark 3.3. By a similar method one can handle the following problem. Let us consider again a clamped beam the deflection of which is limited from below by a rigid obstacle described by a function q. Let $[A, B] \subset (0, 1)$. Our objective is to find a minimal force u under which the beam meets the obstacle on $[A, B]$. In this case

$$V = H_0^2((0, 1)), \quad U = L^2((0, 1)),$$
$$K = \{v \in V \mid v \geq q \text{ a.e. in } (0, 1) \text{ and } v = q \text{ a.e. in } (A, B)\},$$
$$U_{\text{ad}} = \{u \in L^\infty((0, 1)) \mid \alpha \leq u(s) \leq \beta \text{ a.e. in } (0, 1)\},$$

with the given positive constants α and β,

$$\varphi : V \rightarrow (-\infty, +\infty] \text{ is defined by}$$
$$\varphi(y) = \begin{cases} 0 & y \geq q \text{ a.e. in } (0, 1) \\ +\infty & \text{otherwise} \end{cases}$$
$$A(u)y = y^{(iv)}$$
$$B : U \rightarrow H^{-2}((0, 1)) \text{ is the natural embedding mapping.}$$

The state problem admits the form $(\mathbf{S}_3(u))$. The optimal control problem reads

$$(\mathbf{P}_4) \qquad \underset{u \in U_{\text{ad}}}{\text{minimize}} \left\{ J(u) \equiv \int_0^1 u(s) \, ds \right\},$$

where $y(u) \in K$ solves the state problem $(\mathbf{S}_3(u))$.

Remark 3.4. Instead of the penalty technique one could also apply the Lagrange multiplier technique.

We have applied nonsmooth optimization technique to relative simple optimal control problems, because our purpose has only been to demonstrate the use of the method. In next chapters more complicated problems will be handled. For further study of optimal control problems with various sources of nonsmoothness see for example **Barbu** (1984), **Banichuk** (1990), **Bermudez** (1988), **Frankowska** (1984), **Hartl** (1986), **Hlaváček, Bock and Lovíšek** (1984), **Kaškosz and Lojasiewicz** (1988), **Lurie** (1988) (composite materials), **Malanovski** (1987), **Raĭtum** (1979), **Mäkelä** (1990 b), **Myslinski** (1985), **Myslinski and Sokolowski** (1985), **Polyak** (1988) **Roubiček** (1989), **Sokolowski** (1985, 1988), **Teo and Goh** (1988) and **Tiba** (1990).

Chapter 4

Optimal Shape Design

4.1. Introduction

The primary problem often facing designers of structural systems is determining the shape of the structure. Despite the graphical work stations and modern software for analyzing the structure, finding the best geometry for the stucture by "trial and error" is still a very tedious and time consuming task. The goal in optimal shape design (structural optimization, or redesign) is to computerize the design process and therefore shorten the time it takes to design new products or improve the existing design. Structural optimization is widely used in certain applications in the automobile, marine, aerospace industries and in designing truss and shell structures (with minimum weights). In general, however, the structural optimization has just begun to penetrate the industrial community. The design process can be computerized by the integrated FEM (Finite Element Method) and CAD (Computer Aided Design) technologies within the optimization loop.

In Figure 4.1 (a) we see the traditional CAD/FEM–system and in Figure 4.1 (b) the new generation of CAD/FEM–system.

Parameters chosen to describe the design (geometry) of the system are called design variables. The design parameters can either be finite dimensional (vector) or distributed parameters. Optimal shape design problems can be divided roughly into three classes: domain optimization, optimal sizing, and topology optimization.

In domain optimization (or boundary shape optimization) the shape of the domain is sought. Usually the problem is reduced to finding a vector function which defines the unknown boundary, or in discretized case the goal is to find a vector, which defines the position of the design nodes determining the optimal shape for the boundary. In Figure 4.2 (a)–(b) we see two quite different domain optimization problems: finding an optimal shape of an airfoil and optimization of the contact surface of an elastic body on rigid foundation, cf. **Pironneau** (1984), **Haslinger and Neittaanmäki** (1988).

Optimal sizing type of problems were discussed in Chapter 3. In those problems we assume that the layout of the structure is given and we try to find optimal sizes of the structural members. The sizes of the members are chosen as the design parameters that can be of a vector or distributed type. In Figure 4.3 (a)–

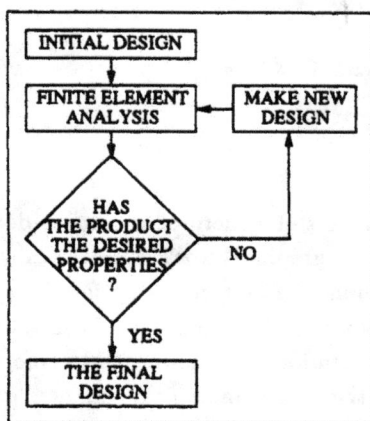
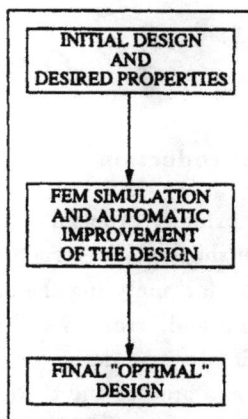

Shape design in traditional CAD system   Optimal shape design system

(a)   (b)

Figure 4.1.

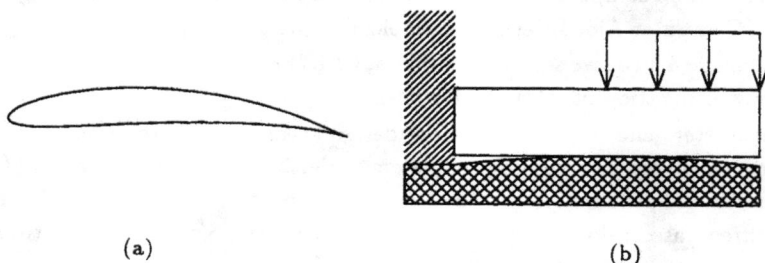

(a)   (b)

Figure 4.2.

(b) we see two typical sizing problems in structural optimization: optimal sizing of a beam (distributed parameter) and of a frame (vector parameter). For further reading we refer to **Arora** (1989), **Brandt** (1986), **Haftka, Gürdal and Kamat** (1990), **Haslinger and Neittaanmäki** (1988), **Haug and Céa** (1981), **Mota Soares** (1987) and references therein.

Topology optimization deals with the search of optimal layout of the system. In Figure 4.4 (a)–(b) we see two simple examples. In the first one an optimal layout of a truss structure is sought. The problem is to find which ones of the dashed members of the truss should exist such that the weight of the truss is minimized and the truss can carry a given load without collapsing. In the second example

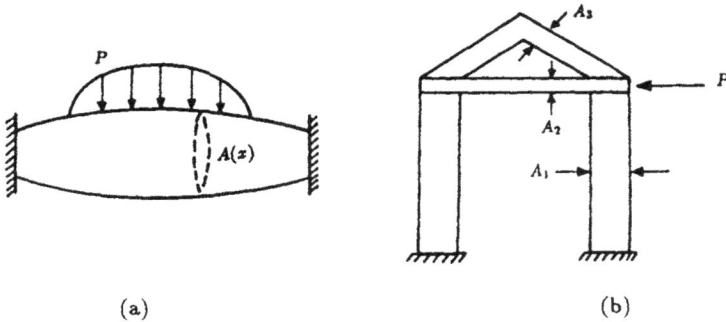

Figure 4.3.

the minimum amount of material should be distributed inside the area surrounded by the dashed curve in such a way that the structure is able to perform some mechanical task. The shaded areas are required to contain material. Topology optimization problems have an on-off nature and are therefore extremely difficult to solve in the distributed case. For various methods of topology optimization see **Kohn and Strang** (1986), **Bendsøe and Kikuchi** (1988) and **Bendsøe and Rodrigues** (1989).

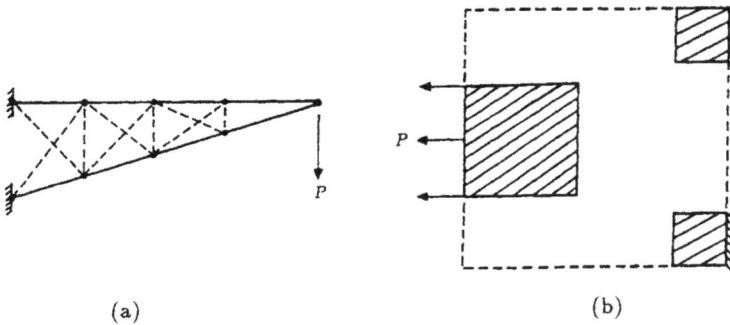

Figure 4.4.

The previous division of optimal shape design problems into three classes is anything but strict. Moreover, it should be noted that the nature of the optimal shape design problem also depends on how the state of the system is modelled.

For example, if the beam in Figure 4.2 (a) is modelled using the classical beam theory the problem is a sizing problem, but if the plane stress model is used the problem is a domain optimization problem. Sometimes it is reasonable to use integrated topology and domain optimization technique, see for example **Bendsøe and Rodrigues** (1989).

This chapter is organized as follows. In Section 4.2 we shall formulate the domain optimization problem in an abstract setting. Several industrial applications which fit into this setting can be found in **Haslinger and Neittaanmäki** (1988) and in **Neittaanmäki** (1991). We shall give here some examples which illustrate the difficulties in the practical numerical realization.

In Section 4.3 we study domain optimization problems where the state system is governed by Dirichlet–Signorini boundary value problem and in Section 4.4 the so called packaging problem is considered. Both problems lead to nonsmooth optimization. We shall present in details the numerical realization, which is based on finite element method and nonsmooth optimization. We shall also compare the performance of the proximal bundle method with the results obtained by SQP method applied to the regularized problem.

4.2. Setting of the Abstract Problem

Let $\Omega \in \mathcal{O}$ (= set of admissible domains) be a domain for which we want to find an optimal design (an optimal geometrical layout). We suppose that $\mathcal{O}$ is a subset of some larger family $\tilde{\mathcal{O}}$; $\mathcal{O} \subseteq \tilde{\mathcal{O}}$.

With any $\Omega \in \tilde{\mathcal{O}}$ we associate a Hilbert space $V(\Omega)$ of functions, defined on Ω. In order to handle the situation mathematically, we introduce a topology in $\tilde{\mathcal{O}}$ and in $\{V(\Omega) \mid \Omega \in \tilde{\mathcal{O}}\}$. If Ω_n, $\Omega \in \tilde{\mathcal{O}}$, we have to define what it means that

$$\Omega_n \xrightarrow{\tilde{\mathcal{O}}} \Omega. \tag{4.1}$$

Analogously, if $y_n \in V(\Omega_n)$, $y \in V(\Omega)$, Ω_n, $\Omega \in \tilde{\mathcal{O}}$, then we must specify the convergence

$$y_n \to y. \tag{4.2}$$

Let

(P) $\qquad\qquad\qquad \Omega \in \mathcal{O} \to y(\Omega) \in V(\Omega)$

be a mapping which with any domain $\Omega \in \mathcal{O}$ associates the solution y of a state problem (given by equations, inequalities etc. in Ω) and let

$$Gr = \{(\Omega, y(\Omega)) \mid \Omega \in \mathcal{O}\} \tag{4.3}$$

be its graph.

Finally, let $I(\Omega, y)$ with $\Omega \in \mathcal{O}$, and $y \in V(\Omega)$ be a cost function (criterion function), whose restriction on Gr will be denoted by $J(\Omega)$, i.e.

$$J(\Omega) = I(\Omega, y(\Omega)). \tag{4.4}$$

The *abstract optimal shape design problem* is stated as follows:

$$\begin{cases} \text{Find } \Omega^* \in \mathcal{O} \text{ such that} \\ J(\Omega^*) \le J(\Omega) \text{ for all } \Omega \in \mathcal{O}. \end{cases} \tag{4.5}$$

We will say that $(\Omega^*, y(\Omega^*))$ is an optimal pair for (4.5).

Theorem 4.1. *Assume that Gr is compact in the following sense: If $\{\Omega_n\}$, $\Omega_n \in \mathcal{O}$ is an arbitrary sequence, there exist a subsequence $\{(\Omega_{n_k}, y(\Omega_{n_k}))\} \subset \{(\Omega_n, y(\Omega_n))\}$ and an element $(\Omega, y(\Omega)) \in Gr$ such that[1] $\Omega_{n_k} \xrightarrow{\mathcal{O}} \Omega$ and $y(\Omega_{n_k}) \to y(\Omega)$.*

Moreover, let I be lower semicontinuous: If Ω_n, $\Omega \in \mathcal{O}$ with $\Omega_n \xrightarrow{\mathcal{O}} \Omega$ and if $y_n \in V(\Omega_n)$, $y \in V(\Omega)$ with $y_n \to y$, then

$$\liminf_{n \to \infty} I(\Omega_n, y_n) \ge I(\Omega, y). \tag{4.6}$$

Then there exists at least one solution $\Omega^* \in \mathcal{O}$ of (4.5).

Proof. See **Haslinger and Neittaanmäki** (1988), Chapter 2. □

A large range of important optimal shape design (domain optimization) problems which arise in structural mechanics, acoustics, electric fields, fluid flow and other areas of engineering and applied sciences can be formulated as problem (4.5). Typically $J(\Omega)$ is

$$\int_\Omega d\Omega \qquad \text{(minimization of the weight)},$$

$$\int_\Omega (y(\Omega))^2 \, d\Omega \quad \text{(minimization of displacements, maximization of stiffness)},$$

$$\int_\Omega (\nabla y(\Omega))^2 \, d\Omega \quad \text{(minimization of energy of deformations)},$$

$$\int_\Gamma \frac{\partial}{\partial n} y(\Omega) \, d\Gamma \quad \text{(minimization of the integral of contact stresses or boundary flux)}.$$

---

[1] Topology in $\mathcal{O}$ is induced by the topology in $\tilde{\mathcal{O}}$.

In the sequel we suppose that the domain Ω can be controlled by a parameter function $u \in U_{ad}$, i.e. $\Omega = \Omega(u)$. Furthermore, we suppose that the boundary Γ of the domain consists of two parts (see Figure 4.5), i.e. $\Gamma = \overline{\Gamma_0} \cup \overline{\Gamma(u)}$, where Γ_0 is the fixed part and $\Gamma(u)$ is the part of the boundary which should be re-designed (optimized) under certain criteria. We solve in $\Omega(u)$ the state problem (partial differential equation, variational inequality, etc.) and obtain the state $y(u)$. Summing up the previous, we have the mappings

$$u \mapsto \Omega(u) \mapsto y(u) \mapsto J(u, y(u))$$

and settle the optimal shape design problem:

(P) $$\underset{u \in U_{ad}}{\text{minimize}} \ J(u, y(u)).$$

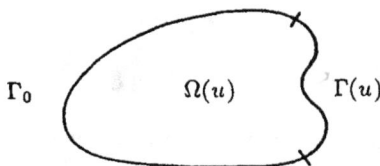

Figure 4.5.

Example 4.1 below shows that in general the mapping $u \mapsto J(u, y(u))$ is not convex and usual optimization methods do not necessarily converge to the optimal solution.

Example 4.1. (Céa (1981)) Let $\Omega(u)$ be an interval of the real line

$$\Omega(u) = \{s \in \mathbf{R} \mid 0 < s < u\}.$$

Let the governing state equation be

$$\begin{cases} -y''(u;s) = 2, & s \in (0, u) \\ y'(u;0) = 0, \\ y(u;u) = 0. \end{cases} \tag{4.7}$$

The cost function to be minimized is

$$J(u) = \int_{\Omega(u)} (y(u) - 1)^2 \, ds. \tag{4.8}$$

As the solution of (4.7) is

$$y(u; s) = u^2 - s^2,$$

the cost function reads by (4.8)

$$J(u) = \frac{8}{15} u^5 - \frac{4}{3} u^3 + u.$$

The graph of J is given in Figure 4.6.

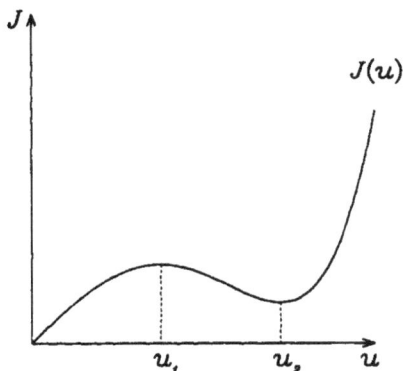

Figure 4.6.

Example 4.2. This example shows a typical trouble, when we are dealing with optimal shape design problem with state system governed by unilateral boundary value problems (variational inequality). Although the original state mapping $u \mapsto y(u)$ is differentiable, it may happen that after discretization the corresponding mapping $x \mapsto y(x)$ is not necessarily differentiable.

Let us consider the problem:

$$\begin{cases} -y''(u; s) \geq -1, & s \in (0, u) \\ y(u; s) \geq 0, \quad y(u; s)(-y''(u; s) + 1) = 0, & s \in (0, u) \\ y(u; 0) = 0, \; y(u; u) = 1. \end{cases} \tag{4.9}$$

The solution of (4.9) is

$$y(u;s) = \left(\frac{1}{u} - \frac{u}{2}\right)s + \frac{1}{2}s^2 \quad \text{for } u \leq \sqrt{2}$$

$$y(u;s) = \begin{cases} \frac{1}{2}(s - u + \sqrt{2})^2 & , \text{if } s \geq u - \sqrt{2} \\ 0 & , \text{if } s \leq u - \sqrt{2} \end{cases} \quad \text{for } u \geq \sqrt{2}.$$

The derivative of y with respect to the design parameter u is

$$\frac{\partial}{\partial u}y(u;s) = \left(-\frac{1}{u^2} - \frac{1}{2}\right)s \quad \text{for } u \leq \sqrt{2}$$

$$\frac{\partial}{\partial u}y(u;s) = \begin{cases} u - s - \sqrt{2} & , \text{if } s \geq u - \sqrt{2} \\ 0 & , \text{if } s \leq u - \sqrt{2} \end{cases} \quad \text{for } u \geq \sqrt{2}.$$

For $u = \sqrt{2}$ both expressions reduce to $-s$. Thus $y(u;s)$ is continuously differentiable in u.

Let $\{0, \frac{x}{3}, \frac{2x}{3}, x\}$ be the partition of $[0, u]$ with $x = u$. The discrete state inequality reads:

$$Ay \geq G(x), \quad y \geq 0, \quad y^T(Ay - G(x)) = 0,$$

where

$$A = \begin{bmatrix} 2 & -1 \\ -1 & 2 \end{bmatrix}, \quad y = \begin{bmatrix} y_1 \\ y_2 \end{bmatrix} \text{ and } G(x) = \begin{bmatrix} 0 \\ -1 \end{bmatrix} - \frac{x^2}{9}\begin{bmatrix} 1 \\ 1 \end{bmatrix}.$$

For $x \leq \sqrt{3}$ we have

$$y_1 = \frac{1}{3}\left(1 - \frac{x^2}{3}\right),$$

$$y_2 = \frac{1}{3}\left(2 - \frac{x^2}{3}\right),$$

i.e. the contact condition $y_i \geq 0$ is not active.

For $x \geq \sqrt{3}$ we have

$$y_1 = 0,$$

$$y_2 = \frac{1}{2}\left(1 - \frac{x^2}{9}\right).$$

Thus it can be seen that neither $y_1(x)$ nor $y_2(x)$ are differentiable in x at $x = \sqrt{3}$ (see Figure 4.7).

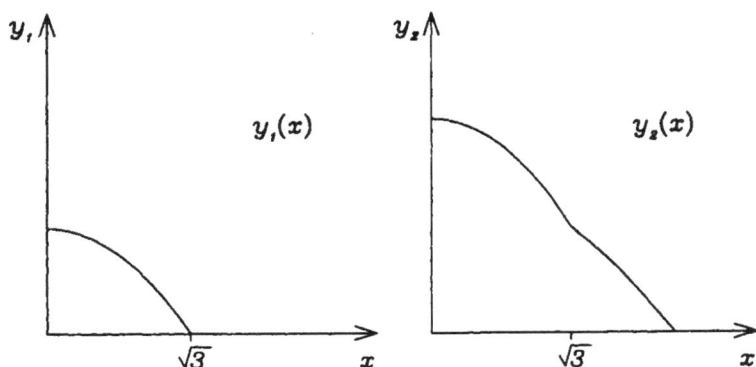

Figure 4.7.

To overcome the difficulty that the state mapping is not differentiable one can apply a *regularization technique* to (**P**) in order to obtain a smooth problem. The regularization technique (exterior penalty technique) is applied for optimal shape design problems governed by unilateral boundary value problems (Dirichlet–Signorini boundary value problems in elasticity and plasticity, obstacle problems etc.) in **Haslinger and Neittaanmäki** (1988). In **Neittaanmäki and Stachurski** (1991) a new technique for regularizing the state problem is presented. The technique is based on a detailed analysis of the nature of nonsmoothness of the mapping $x \mapsto y(x)$. When necessary the solution is slightly modified by a barrier function type technique. This technique enables us to use standard algorithms of nonlinear programming developed for smooth problems. The barrier function regularization technique is natural because the state constraints mean technological constraints and their violation is not favourable. In sequel we introduce the direct approach without any regularization and apply the methods of nonsmooth optimization.

4.3. Optimal Shape Design with Unilateral Boundary Value Problem

In this section we study domain optimization problems where the state system is governed by Dirichlet–Signorini boundary value problems. This problem was introduced and handled using variational inequality and exterior penalty method in **Haslinger and Neittaanmäki** (1988).

4.3.1. Setting of the Problem

We suppose that the geometrical situation is as described in Figure 4.8, i.e.

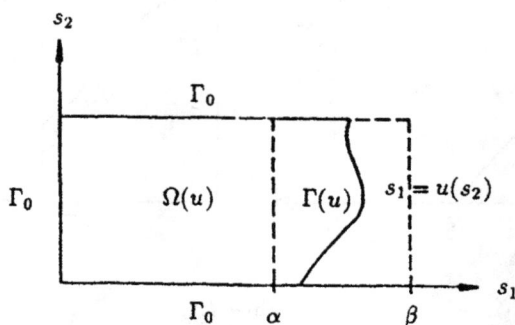

Figure 4.8.

$$\Omega(u) = \{(s_1, s_2) \in \mathbf{R}^2 \mid 0 < s_1 < u(s_2),\ s_2 \in (0,1),\ u \in U_{\mathrm{ad}}\},$$

where the set of admissible controls U_{ad} is defined by

$$U_{\mathrm{ad}} = \{u \in C^{0,1}([0,1]) \mid 0 < \alpha \leq u(s_2) \leq \beta,$$

$$|u'(s_2)| \leq \gamma \quad \text{for a.a. } s_2 \in [0,1],\ \text{meas } \Omega(u) = M\} \quad (4.10)$$

with given positive constants α, β, γ and M such that $U_{\mathrm{ad}} \neq \emptyset$. Here meas $\Omega(u) = \int_0^1 u(s_2)\, ds_2$ denotes the measure of the domain $\Omega(u)$. The boundary Γ of $\Omega(u)$ is given by $\overline{\Gamma}_0 \cup \overline{\Gamma(u)}$, where the part of the boundary to be redesigned is parametrized as follows

$$\Gamma(u) = \{(s_1, s_2) \in \mathbf{R}^2 \mid s_1 = u(s_2),\ s_2 \in (0,1)\}.$$

Suppose that $y(u)$ is governed by a scalar Dirichlet–Signorini boundary value problem

$$\begin{cases} -\Delta y(u) = g & \text{in } \Omega(u) \\ y(u) = 0 & \text{on } \Gamma_0 \\ y(u) \geq 0,\ \dfrac{\partial}{\partial n} y(u) \geq 0,\ y(u)\dfrac{\partial}{\partial n} y(u) = 0 & \text{on } \Gamma(u), \end{cases} \quad (4.11)$$

where $g \in L^2(\hat{\Omega})$, $\hat{\Omega} = (0,1) \times (0,\hat{\beta})$, $\hat{\beta} > \beta$, $u \in U_{ad}$ and $\frac{\partial}{\partial n}y = n^T \nabla y$ denotes the outward normal derivative of y. Here $y(u)$ is the vertical displacement and g is a given vertical force. For convenience we suppose in design sensitivity analysis that g in (4.11) is differentiable.

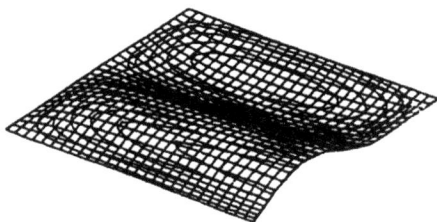

Figure 4.9.

On $\Gamma(u)$ we see the nature of the Signorini type boundary conditions. We note that it is not a priori known where the various boundary conditions $y = 0$, $y > 0$, $\frac{\partial}{\partial n}y = 0$ or $\frac{\partial}{\partial n}y > 0$ are realized. We only know that $y\frac{\partial}{\partial n}y = 0$ on $\Gamma(u)$, which implies that if $y > 0$ then $\frac{\partial}{\partial n}y = 0$. This can be seen in Figure 4.9, where the spline smoothened finite element solution of (4.11) is illustrated in the case where $\Omega = (0,1) \times (0,1)$, i.e. $u \equiv 1$ and $g(s_1, s_2) = 4\sin 2\pi s_2$.

We consider the following three cost functionals, with respect to which the goodness of the design will be controlled, namely

$$
\begin{cases}
J_1(u, y(u)) = \frac{1}{2} \int\limits_{\Omega(u)} y(u)^2 \, d\Omega(u) \\[2ex]
J_2(u, y(u)) = \frac{1}{2} \int\limits_{\Gamma(u)} y(u)^2 \, d\Gamma(u) \\[2ex]
J_3(u, y(u)) = \int\limits_{\Gamma(u)} \frac{\partial}{\partial n}y(u)\Phi \, d\Gamma(u),
\end{cases}
\tag{4.12}
$$

where $\Phi \in M(\hat{\Omega})$ is a non-negative function,

$$M(\hat{\Omega}) = \{\Phi \in H_0^1(\hat{\Omega}) \mid \Phi \geq 0 \text{ in } \hat{\Omega}, \ \frac{\partial}{\partial s_1}\Phi = 0 \text{ a.e. in } (\alpha, \beta) \times (0, 1), \ \|\Phi\|_{1,\hat{\Omega}} \leq 1\}.$$

Our optimal shape design problem reads now

$(\mathbf{P}_l)$ $$\underset{u \in U_{ad}}{\text{minimize}} \ J_l(u, y(u)) \qquad l = 1, 2, 3,$$

subject to $y = y(u) \in K$ solving the variational inequality

$(\mathbf{S}(u))$ $$a_u(y(u), v - y(u)) \geq (g, v - y(u))_{0,\Omega(u)} \quad \text{for all} \quad v \in K,$$

where

$$K = \{v \in H^1(\Omega(u)) \mid v|_{\Gamma_0} = 0 \ \text{ and } v \geq 0 \text{ on } \Gamma(u)\}$$

and

$$a_u(y, v) = \int_{\Omega(u)} \nabla y \nabla v \ d\Omega(u)$$

$$(g, v)_{0,\Omega(u)} = \int_{\Omega(u)} gv \ d\Omega(u).$$

Concerning the solvability of $(\mathbf{P}_l)$ we have the following result.

Theorem 4.2. *Let $U_{ad} \neq \emptyset$. Then there exists at least one solution of $(\mathbf{P}_l)$.*

Proof. The assumptions of Theorem 4.1 are valid. In this case

$$\mathcal{O} = \{\Omega(\alpha) \mid \alpha \in U_{ad}\}.$$

See **Haslinger and Neittaanmäki** (1988), Chapter 3 for further details. □

Remark 4.1. In problem $(\mathbf{P}_1)$ we minimize the displacement over the whole domain $\Omega(u)$, in $(\mathbf{P}_2)$ over the moving boundary $\Gamma(u)$ and in $(\mathbf{P}_3)$ the flux across the moving boundary. It would also be reasonable to consider all these criteria together and apply the methods of multicriteria optimization (see **Miettinen and Mäkelä** (1991)).

Remark 4.2. By partial integration J_3 can be transformed as functional defined in the domain by

$$J_{3,\Phi}(u, y(u)) = \int_{\Omega(u)} \nabla y(u) \nabla \Phi \ d\Omega(u) - \int_{\Omega(u)} f\Phi \ d\Omega(u). \qquad (4.13)$$

This form of J_3 enables us to calculate more accurate numerical values than in the original formulation.

4.3.2. Discretization of the Problem

We shall approximate the infinite dimensional problems (P_l) for $l = 1, 2, 3$ by finite dimensional ones. In this connection only some headlines and basic ideas of the discretization technique are presented. For further details we refer to **Haslinger and Neittaanmäki** (1988).

We shall use linear finite elements for approximating y and u. Let us denote these approximations by y_h and u_h, where $h > 0$ is a fixed discretization parameter. Let $T(h, u_h)$ denote a uniformly regular triangulation of $\overline{\Omega(u_h)}$: $\overline{\Omega(u_h)} = \cup_{T \in T(h, u_h)} T$. We denote the nodes by N_i for $i \in I$, where I is the set of indices of the nodes lying on $\Omega(u_h) \cup \overline{\Gamma(u_h)}$. We split I as follows

$$I = I_{\Omega(u_h)} \cup I_{\overline{\Gamma(u_h)}} = I_{\Omega(u_h)} \cup I_{\Gamma(u_h)} \cup I_{\overline{\Gamma(u_h)} \setminus \Gamma(u_h)}.$$

Further let

$$m = m(h) = \text{card } (I_{\Omega(u_h)} \cup I_{\Gamma(u_h)})$$
$$n = n(h) = \text{card } (I_{\overline{\Gamma(u_h)}})$$

and suppose that the nodes lying on $\Gamma(u_h)$ are numbered last. Thus m is the number of nodes lying on $\Omega(u_h) \cup \Gamma(u_h)$ and n is the number of the design nodes N_i^D lying on $\overline{\Gamma(u_h)}$. We suppose that $0 = a_1 < a_2 < \cdots < a_n = 1$ is an equidistant partition of $[0, 1]$, thus $a_i = (i - 1)h$ for $i = 1, \ldots, n$; i.e. $n = n(h) = (1 + h)/h$. Further we suppose that $u_h \in U_{\text{ad}}^h$, where

$$U_{\text{ad}}^h = \{u_h \mid u_h|_{[a_{i-1}, a_i]} \in P_1, \ i = 2, \ldots, n, \ 0 < \alpha \le u_h(s_2) \le \beta,$$
$$|u_h(s_2) - u_h(s_2')| \le \gamma|s_2 - s_2'|, \ s_2, s_2' \in [0, 1], \ \text{meas } \Omega(u_h) = M\} \ .$$

Because α_h is piecewise linear the shape of $\Gamma(u_h)$ (and hence of $\Omega_h(u_h)$) is uniquely determined by the s_1-coordinates of the design nodes $N_i^D = (u_h(a_i), a_i)$, $i = 1, \ldots, n$ (see Figure 4.10). The triangulation $T(h, u_h)$ is parametrized in such a way that it depends continuously on u_h and is supposed to satisfy the standard requirements for a regular triangulation (see **Križek and Neittaanmäki** (1990), for example); $T(h, u_h)$ consists of two parts: a fixed triangulation of $\overline{\Omega(\alpha')} = [0, \alpha'] \times [0, 1]$ with $\alpha' < \alpha$ and a moving triangulation of $\overline{\Omega(u_h) \setminus \Omega(\alpha')}$. The moving triangulation of $\overline{\Omega(u_h) \setminus \Omega(\alpha')}$ can be constructed for example by defining an equidistant partition $N_{i,1}^M = \hat{N}_i < N_{i,2}^M < \cdots < N_{i,k}^M = N_i^D$ of interval $[\hat{N}_i, N_i^D]$ (see Figure 4.10). For convenience we suppose that the nodes N_i which are design nodes are listed last in numbering the nodes of $\overline{\Omega(u_h)}$.

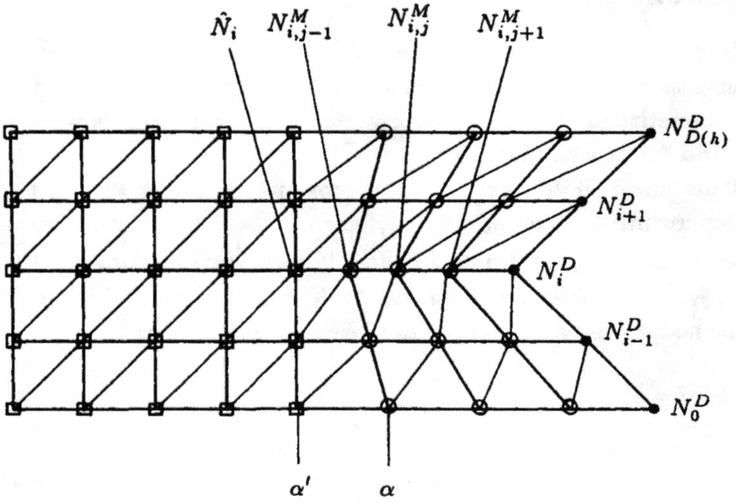

Figure 4.10.

The discretized variants of Problems $(\mathbf{P}_l)$, $l = 1, 2, 3$ read

$$(\mathbf{P}_{l,h}) \qquad\qquad \underset{u_h \in U_{\mathrm{ad}}^h}{\text{minimize}}\ J_{l,h}(u_h, y_h)$$

subject to $y_h = y_h(u_h) \in K_h$ solving the variational inequality

$$(\mathbf{S}_h(u_h)) \quad a_{u_h}(y_h(u_h), v_h - y_h(u_h)) \geq (g, v_h - y_h(u_h))_{0,\Omega(u_h)} \quad \text{for all } v_h \in K_h,$$

where

$$K_h = \{v_h \in C(\overline{\Omega(u_h)}) \mid v_h\big|_T \in P_1\ \forall\ T \in \mathcal{T}(h, u_h),$$
$$v_h = 0 \text{ on } \Gamma_0 \text{ and } v_h \geq 0 \text{ on } \Gamma(u_h)\}.$$

In $(\mathbf{P}_{l,h})$: $J_{3,h}(u_h, y_h) = J_{3,\Phi}(u_h, y_h(u_h))$ (see (4.13)).

In order to present Problem $(\mathbf{P}_{l,h})$ in an algebraic form we first note that the set U_{ad}^h can be identified with a closed convex subset of $\mathbf{R}^n$ given by

$$\mathcal{U} = \{x \in \mathbf{R}^n \mid 0 < \alpha \leq x_i \leq \beta,\ i = 1, \ldots, n,$$
$$-\gamma h \leq x_{i+1} - x_i \leq \gamma h\ i = 1, \ldots, n-1,\ \sum_{i=1}^{n-1}(x_{i+1} + x_i) = 2M/h\}. \qquad (4.14)$$

We define the "discrete design vector"

$$x = (x_1, \ldots, x_n)^{\mathrm{T}},$$

where the design (or control) variables are $x_i = u_h(a_i)$ for $i = 1, \ldots, n$. Let

$$y_h(s) = \sum_{i=1}^{m} y_i(x)\varphi_i(s),$$

where φ_i is the Courant basis function with $\varphi_i(N_j) = \delta_{ij}$ and $y \in \mathbf{R}^m$ is the vector of nodal values of FEM-solution y_h, that is

$$y_i = y_h(N_i), \qquad i = 1, \ldots, m.$$

Further, let $\phi \in \mathbf{R}^m$ denote the nodal value vector of Φ, i.e.

$$\phi_i = \Phi(N_i), \qquad i = 1, \ldots, m.$$

Because of the unilateral boundary condition we have $y \in \mathcal{K}$, where

$$\mathcal{K} = \{y \in \mathbf{R}^m \mid y_i \geq 0 \quad \forall i \in I_{\Gamma(u_h)}\}.$$

Let

$$A(x) = \left(A_{ij}(x)\right)_{i,j=1}^{m} = \left(\int_{\Omega(u_h)} \nabla\varphi_i \nabla\varphi_j \, d\Omega(u_h)\right)_{i,j=1}^{m}$$

be the stiffness matrix,

$$M_1(x) = \left(M_{1,ij}(x)\right)_{i,j=1}^{m} = \left(\int_{\Omega(u_h)} \varphi_i\varphi_j \, d\Omega(u_h)\right)_{i,j=1}^{m}$$

and

$$\hat{M}_2(x) = \left(\hat{M}_{2,ij}(x)\right)_{i,j=m-n}^{m} = \left(\int_{\Gamma(u_h)} \varphi_i|_{\Gamma(u_h)} \, \varphi_j|_{\Gamma(u_h)} \, d\Gamma(u_h)\right)_{i,j=m-n}^{m}$$

the mass matrices and

$$G(x) = \left(G_j(x)\right)_{j=1}^{m} = \left(\int_{\Omega(u_h)} g\varphi_j \, d\Omega(u_h)\right)_{j=1}^{m}$$

the force vector. The matrix $A(x)$ is symmetric and positive definite uniformly with respect to $x \in \mathcal{U}$. Through the parametrization of the triangulation, the mappings $x \mapsto A(x)$, $x \mapsto M_l(x)$ and $x \mapsto G(x)$ are continuously differentiable (of course supposing that g is differentiable).

The algebraic form of the state problem $(\mathbf{S}_h(u_h))$ for a fixed $x \in \mathcal{U}$ can be formulated as follows: Find $y = y(x) \in \mathcal{K}$ such that

$$(\mathcal{S}(x)) \qquad (v - y(x))^{\mathrm{T}} A(x) y(x) \geq G(x)^{\mathrm{T}} (v - y(x)) \quad \text{for all } v \in \mathcal{K}$$

or equivalently: Find $y = y(x) \in \mathcal{K}$ such that

$$(\mathcal{S}(x)') \qquad y = \arg\min_{z \in \mathcal{K}} \left\{ \tfrac{1}{2} z^{\mathrm{T}} A(x) z - G(x)^{\mathrm{T}} z \right\}.$$

The discrete matrix forms of cost functionals (4.12) read now

$$\begin{cases} \mathcal{J}_1(x, y(x)) = \tfrac{1}{2} y(x)^{\mathrm{T}} M_1(x) y(x) \\ \mathcal{J}_2(x, y(x)) = \tfrac{1}{2} y(x)^{\mathrm{T}} M_2(x) y(x) \\ \mathcal{J}_3(x, y(x)) = \mathcal{J}_{3,\phi}(x, y(x)) = y(x)^{\mathrm{T}} K(x) \phi - G(x)^{\mathrm{T}} \phi, \end{cases} \qquad (4.15)$$

where

$$M_2 = \begin{pmatrix} 0 & 0 \\ 0 & \hat{M}_2 \end{pmatrix}.$$

The discrete forms of the problems $(\mathbf{P}_l)$ for $l = 1, 2, 3$ take the form

$$(\mathcal{P}_l) \qquad\qquad \min_{x \in \mathcal{U}} \mathcal{J}_l(x, y(x))$$

subject to $y(x)$ solving the state problem $(\mathcal{S}(x))$.

As $\mathcal{U} \subset \mathbf{R}^n$ is compact and as the mapping $x \mapsto \mathcal{J}_l(x, y(x))$ is continuous (Theorem 4.3), Problem (P_l) has at least one solution.

Remark 4.3. Problems $(\mathcal{P}_l)$ for $l = 1, 2, 3$ are nonlinear optimization problems with linear constraints and simple bounds for the variables. The difficulties in the numerical realization of $(\mathcal{P}_l)$ are:

- The evaluation of the function $\mathcal{J}_l$ is costly. Every function evaluation means the solving of the nonlinear state system $(\mathcal{S}(x))$.
- As the mapping $x \mapsto y(x)$ is only Lipschitz continuous (see Theorem 4.3), the mapping $x \mapsto \mathcal{J}_l(x, y(x))$ is not differentiable. Moreover, the mapping $x \mapsto \mathcal{J}_l(x, y(x))$ is not necessarily convex (see Example 4.1). Consequently, the $(\mathcal{P}_l)$ are nonsmooth, nonconvex optimization problems.

Theorem 4.3. *The mapping* $x \mapsto y(x)$ *is Lipschitz continuous.*

Proof.

As the triangulation $T(h, u_h)$ depends continuously on u_h, the mappings $x \mapsto$ $A(x)$ and $x \mapsto G(x)$ are continuous. It can be shown that these mappings in fact are differentiable (**Haslinger and Neittaanmäki** (1988), Appendix II).

For x, $\bar{x} \in \mathcal{U}$ we have

$$
\begin{aligned}
y(x) \in \mathcal{K} : \ (z - y(x))^T A(x) y(x) \geq G(x)^T (z - y(x)) \quad \forall z \in \mathcal{K}, \\
y(\bar{x}) \in \mathcal{K} : \ (z - y(\bar{x}))^T A(\bar{x}) y(\bar{x}) \geq G(\bar{x})^T (z - y(\bar{x})) \quad \forall z \in \mathcal{K}.
\end{aligned}
\tag{4.16}
$$

Substituting $z = y(\bar{x})$ in the first inequality and $z = y(x)$ in the second and adding them together we get

$$
(y(\bar{x}) - y(x))^T (A(x) y(x) - A(\bar{x}) y(\bar{x})) \geq (G(x) - G(\bar{x}))^T (y(\bar{x}) - y(x)). \tag{4.17}
$$

Adding and subtracting $(y(\bar{x}) - y(x))^T A(x) y(\bar{x})$, and finally rearranging terms in (4.17) we get

$$
\begin{aligned}
(y(\bar{x}) &- y(x))^T A(x) (y(\bar{x}) - y(x)) \\
&\leq (y(\bar{x}) - y(x))^T (A(x) - A(\bar{x})) y(\bar{x}) + (G(x) - G(\bar{x}))^T (y(x) - y(\bar{x})) \\
&\leq \|A(x) - A(\bar{x})\| \, \|y(\bar{x})\| \, \|y(\bar{x}) - y(x)\| \\
&\quad + \|G(x) - G(\bar{x})\| \, \|y(x) - y(\bar{x})\| \\
&\leq C \|y(\bar{x}) - y(x)\| \{ \|A(x) - A(\bar{x})\| + \|G(x) - G(\bar{x})\| \}
\end{aligned}
\tag{4.18}
$$

making use of the uniform boundedness of $y(x)$, $x \in \mathcal{U}$.

As $A(x)$ is positively definite, uniformly with respect to $x \in \mathcal{U}$, we finally get

$$
\|y(x) - y(\bar{x})\| \leq C \{ \|A(x) - A(\bar{x})\| + \|G(x) - G(\bar{x})\| \}, \tag{4.19}
$$

from which the assertion follows by the continuity of the mappings $x \mapsto A(x)$ and $x \mapsto G(x)$. $\qquad \square$

4.3.3. Design Sensitivity Analysis

In the sensitivity analysis we can again apply Theorem 2.1 in order to calculate one subgradient from the subdifferential $\partial \mathcal{J}_l$. Before doing this we introduce some notations.

Let $y(x) = (y_1(x), \ldots, y_m(x)) \in \mathbf{R}^m$, $x = (x_1, \ldots, x_n) \in \mathbf{R}^n$. Let $M(x) = (M_{ij}(x))_{i,j=1}^m \in \mathbf{R}^{m \times m}$ be a matrix depending on $x \in \mathbf{R}^n$. For a multilinear mapping $\nabla M(x)$ we denote

$$\nabla M(x)[y] = \left(\sum_{k=1}^m \frac{\partial}{\partial x_j} M_{ik}(x) y_k \right)_{i,j=1}^{m,n} \in \mathbf{R}^{m \times n},$$

$$\nabla M(x)[y, v] = v^{\mathrm{T}} \nabla M(x)[y] \in \mathbf{R}^n.$$

Theorem 4.4. *Let* $\mathcal{J}_l : \mathbf{R}^n \to \mathbf{R}$ *be defined by (4.15),* $l = 1, 2, 3$. *If* $\xi_y(x) \in \partial y(x)$, *then for* $l = 1, 2$

$$\tfrac{1}{2} \nabla M_l(x)[y(x), y(x)] + \xi_y(x)^{\mathrm{T}} M_l(x) y(x) \in \partial \mathcal{J}_l(x, y(x)) \qquad (4.20)$$

and

$$(\nabla A(x)[y(x)] - \nabla G(x))^{\mathrm{T}} \phi + z_y(x)^{\mathrm{T}} A(x) \phi \in \partial \mathcal{J}_3(x, y(x)). \qquad (4.21)$$

Proof. Follows from Theorem 2.1, as in this case

$$f(x) = \mathcal{J}_l(x, y(x)), \quad l = 1, 2, 3,$$

i.e. $h(x) = 0$ and $E(x, y(x)) = \mathcal{J}_l(x, y(x))$, $l = 1, 2, 3$. Consequently,

$$\nabla h(x) = 0$$

$$\nabla_x E(x, y(x)) = \begin{cases} \tfrac{1}{2} \nabla M_1(x)[y(x), y(x)], & l = 1, \\ \tfrac{1}{2} \nabla M_2(x)[y(x), y(x)], & l = 2, \\ (\nabla A(x)[y(x)] - \nabla G(x))^{\mathrm{T}} \phi, & l = 3 \end{cases} \qquad (4.22)$$

and

$$\nabla_y E(x, y(x)) = \nabla_y \mathcal{J}_l(x, y(x)) = \begin{cases} M_1(x) y(x), & l = 1, \\ M_2(x) y(x), & l = 2, \\ A(x) \phi, & l = 3. \end{cases} \qquad (4.23)$$

$\square$

The computation of the representative $\xi_y(x)$ from the generalized Jacobian $\partial y(x)$ is a troublesome task. In practice one can avoid the explicit performing of this calculus by means of the so called adjoint state technique.

In order to demonstrate the adjoint state technique we *suppose for a moment* that instead of the variational inequality $(S(u))$ the state is given by the variational equation (with Dirichlet-Neumann boundary condition which we obtain if in the Dirichlet-Signorini boundary value problem (4.21) $y(u) > 0$ on $\Gamma(u)$). In this case the state mapping $x \mapsto y(x)$ is differentiable and $(S(x))$ reduces to

$$(\widetilde{S}(x)) \qquad\qquad A(x)y(x) = G(x).$$

A differentiation of $(\widetilde{S}(x))$ gives

$$A(x)\nabla y(x) = \nabla G(x) - \nabla A(x)[y(x)]. \tag{4.24}$$

Let $p_l(x)$, $l = 1, 2, 3$, be the solution of the adjoint state problem

$$(\widetilde{\mathcal{A}}(x)) \qquad\qquad A(x)p_l(x) = \nabla_y \mathcal{J}_l(x, y(x)), \quad l = 1, 2, 3.$$

By (4.20), $(\widetilde{\mathcal{A}}(x))$ and (4.24), $(\xi_y(x) = \nabla y(x)$ in the differentiable case)

$$
\begin{aligned}
\nabla \mathcal{J}_1(x, y(x)) &= \tfrac{1}{2}\nabla M_1(x)[y(x), y(x)] + \nabla y(x)^{\mathrm{T}} \nabla_y \mathcal{J}_1(x, y(x)) \tag{4.25}\\
&= \tfrac{1}{2}\nabla M_1(x)[y(x), y(x)] + \nabla y(x)^{\mathrm{T}} A(x)p_1(x) \\
&= \tfrac{1}{2}\nabla M_1(x)[y(x), y(x)] + (A(x)\nabla y(x))^{\mathrm{T}} p_1(x) \\
&= \tfrac{1}{2}\nabla M_1(x)[y(x), y(x)] + (\nabla G(x) - \nabla A(x)[y(x)])^{\mathrm{T}} p_1(x).
\end{aligned}
$$

Similarly we obtain

$$\nabla \mathcal{J}_2(x, y(x)) = \tfrac{1}{2}\nabla M_2(x)[y(x), y(x)] + (\nabla G(x) - \nabla A(x)[y(x)])^{\mathrm{T}} p_2(x). \tag{4.26}$$

For $\mathcal{J}_3$ we have

$$
\begin{aligned}
\nabla \mathcal{J}_3(x, y(x)) &= (\nabla A(x)[y(x)] - \nabla G(x))^{\mathrm{T}} \phi + (\nabla G(x) - \nabla A(x)[y(x)])^{\mathrm{T}} p_3(x) \\
&= (\nabla A(x)[y(x)] - \nabla G(x))^{\mathrm{T}} (\phi - p_3(x)) \equiv 0,
\end{aligned}
$$
$$\tag{4.27}$$

since due to (4.23) and $(\widetilde{\mathcal{A}}(x))$ we have $p_3 \equiv \phi$. This is natural, because by (4.21) for every $u \in U_{\mathrm{ad}}$ with $y(u) > 0$ implies the Neumann condition $\frac{\partial}{\partial n}y = 0$ and consequently $J_{3,\Phi}(u, y(u)) = 0$.

Let us now consider the adjoint state technique in the case of Dirichlet–Signorini boundary conditions. We shall first analyse the dependence of the solution $y(x)$ of $(S(x))$ on variations of the discrete design variable x. The classical first order Karush–Kuhn–Tucker optimality conditions, applied to $(S(x)')$, guarantee the existence of a non-negative vector of Lagrange multipliers $\lambda = \lambda(x) \in \mathbb{R}^m$ such that

$$\begin{cases} A(x)y(x) = G(x) + \lambda(x) \\ y_i(x), \lambda_i(x) \geq 0, \; y_i(x)\lambda_i(x) = 0 \qquad \text{for all} \quad i \in I_{\Gamma(u_h)} \qquad (4.28) \\ \qquad\qquad\qquad \lambda_i(x) = 0 \qquad \text{for all} \quad i \in I_{\Omega(u_h)}. \end{cases}$$

In order to derive relations for finding the directional derivative $y'(x; d) \in \mathbb{R}^m$ in the direction $d \in \mathbb{R}^n$ we first differentiate both sides of the first equation in (4.28), and we get

$$A(x)y'(x; d) = \nabla G(x)d + \lambda'(x; d) - \nabla A(x)[y(x)]d, \qquad (4.29)$$

where $\lambda'(x; d)$ denotes the directional derivative of the Lagrange multipliers as a function of x. To analyse the situation in (4.29) further we split $I_{\Gamma(u_h)}$ in the following way

$$I_{\Gamma(u_h)} = I_a(x) \cup I_n(x) \cup I_s(x)$$

with

$$I_a(x) = \{i \in I_{\Gamma(u_h)} \mid y_i(x) = 0 \text{ and } \lambda_i(x) > 0\}$$
$$I_n(x) = \{i \in I_{\Gamma(u_h)} \mid y_i(x) > 0 \text{ and } \lambda_i(x) = 0\}$$
$$I_s(x) = \{i \in I_{\Gamma(u_h)} \mid y_i(x) = 0 \text{ and } \lambda_i(x) = 0\}$$

(the sets of active, nonactive constraints and semiactive at x). As $\lambda_i(x + td) > 0$ for $i \in I_a(x)$ and $0 \leq t < t_1$, t_1 sufficiently small, we get $y'_i(x; d) = 0$, $i \in I_a(x)$. Similarly, if $i \in I_n(x)$ then $y_i(x + td) > 0$, $0 \leq t \leq t_2$ and, consequently, $\lambda'_i(x; d) = 0$, $i \in I_n(x)$. If $i \in I_s(x)$, then $\lambda'_i(x; d) \geq 0$, $y'_i(x; d) \geq 0$ and $\lambda'_i(x; d)y'_i(x; d) \geq 0$. Let us prove that $\lambda'_i(x; d)y'_i(x; d) = 0$. If $\lambda'_i(x; d) = 0$, the previous equality holds. If $\lambda'_i(x; d) > 0$, then $\lambda_i(x + td) > 0$ for $t > 0$ sufficiently small. Hence $y_i(x + td) = 0$ and consequently $y'_i(x; d) = 0$. Combining all these relations together we get the following result.

Theorem 4.5. *Let $y(x) \in \mathcal{K}$ be the solution of $(\mathcal{S}(x))$. Then the directional derivative $y'(x; d)$ at $x \in \mathcal{U}$ in the direction $d \in \mathbb{R}^n$ can be found as a solution of the following problem*

$$\begin{cases} (A(x)y'(x; d))_i = (\nabla G(x)d)_i - (\nabla A(x)[y(x)]d)_i, \quad i \in I_{\Omega(u_h)} \cup I_n(x) \\ \qquad\qquad\qquad\qquad\qquad y'_i(x; d) = 0, \quad i \in I_a(x) \\ \qquad\qquad\qquad\qquad\qquad y'_i(x; d) \geq 0, \\ (A(x)y'(x; d))_i - (\nabla G(x)d)_i + (\nabla A(x)[y(x)]d)_i \geq 0, \\ y'_i(x; d)((A(x)y'(x; d))_i - (\nabla G(x)d)_i + (\nabla A(x)[y(x)]d)_i) = 0, \quad i \in I_s(x). \end{cases}$$

Remark 4.4. If $I_s(x) \neq \emptyset$, then, in general, $y'_i(x; d)$ does not depend linearly on d, i.e. $\nabla y(x)$ does not exist.

In order to eliminate the term $\xi_y(x)$ in (4.29) and (4.21) respectively, we introduce (analogously to $(\tilde{\mathcal{A}}(x))$) the adjoint state $p_l(x) \in \mathbf{R}^m$ (when it exists) to Problem $(\mathcal{P}_l)$, $l = 1, 2, 3$ as the solution of the quadratic programming problem

$$(\mathcal{A}(x)) \qquad \min_{p \in \tilde{\mathcal{K}}(x)} \left\{ \tfrac{1}{2} p^{\mathrm{T}} A(x) p - p^{\mathrm{T}} \nabla_y \mathcal{J}_l(x, y(x)) \right\}, \quad l = 1, 2, 3,$$

where

$$\tilde{\mathcal{K}}(x) = \{ p \in \mathbf{R}^m \, | \, p_i = 0 \quad \text{for all} \quad i \in I_a(x)$$
$$p_i \geq 0 \quad \text{for all} \quad i \in I_s(x) \}.$$

Now we can apply the Karush–Kuhn–Tucker optimality conditions to $(\mathcal{A}(x))$ and we obtain a vector of Lagrange multipliers $\mu_l(x) \in \mathbf{R}^m$ such that

$$\left\{ \begin{aligned} A(x) p_l(x) &= \nabla_y \mathcal{J}_l(x, y(x)) + \mu_l(x) & l &= 1, 2, 3, \\ p_{l,i}(x) &= 0 & \text{for all} \quad i &\in I_a(x) \\ \mu_{l,i}(x) &= 0 & \text{for all} \quad i &\in I_{\Omega(u_h)} \cup I_n(x) \\ \mu_{l,i}(x) &\geq 0, p_{l,i}(x) \geq 0, p_{l,i}(x) \mu_{l,i}(x) = 0 & \text{for all} \quad i &\in I_s(x). \end{aligned} \right. \tag{4.30}$$

By using (4.20), (4.23) and (4.30) we obtain

$$\tfrac{1}{2} \nabla M_1(x)[y(x), y(x)] + \xi_y(x)^{\mathrm{T}} (A(x) p_1(x) - \mu_1(x))$$
$$= \tfrac{1}{2} \nabla M_1(x)[y(x), y(x)] + (A(x) \xi_y(x))^{\mathrm{T}} p_1(x) - \xi_y(x)^{\mathrm{T}} \mu_1(x) \in \partial \mathcal{J}_1(x, y(x)).$$

Due to (4.28) we have

$$\partial (A(x) y(x)) = \partial G(x) + \partial \lambda(x),$$

which implies that

$$A(x) \xi_y(x) = \nabla G(x) - \nabla A(x)[y(x)] + \xi_\lambda(x)$$

for some $\xi_\lambda(x) \in \partial \lambda(x)$. Consequently,

$$\tfrac{1}{2} \nabla M_1(x)[y(x), y(x)] + (\nabla G(x) - \nabla A(x)[y(x)] + \xi_\lambda(x))^{\mathrm{T}} p_1(x)$$
$$- \xi_y(x)^{\mathrm{T}} \mu_1(x) \in \partial \mathcal{J}_1(x, y(x)).$$

If $I_s(x) = \emptyset$, then by (4.30) and Theorem 4.5 we have $\xi_\lambda(x)^{\mathrm{T}} p_1(x) = 0$ and $\xi_y(x)^{\mathrm{T}} \mu_1(x) = 0$. Thus we have

$$\nabla \mathcal{J}_1(x, y(x)) = \tfrac{1}{2} \nabla M_1(x)[y(x), y(x)] - (\nabla G(x) - \nabla A(x)[y(x)])^{\mathrm{T}} p_1(x). \quad (4.31)$$

Analogously one can prove

$$\nabla \mathcal{J}_2(x, y(x)) = \tfrac{1}{2} \nabla M_2(x)[y(x), y(x)] + (\nabla G(x) - \nabla A(x)[y(x)])^{\mathrm{T}} p_2(x) \quad (4.32)$$

and

$$\nabla \mathcal{J}_3(x, y(x)) = (\nabla A(x)[y(x)] - \nabla G(x))^{\mathrm{T}} (\phi - p_3(x)), \quad (4.33)$$

where p_2 and p_3 are adjoint states, being the solution of $(\mathcal{A}(x))$ for $i = 1, 2$, respectively. Notice that in this case $p_3 \neq \phi$ due to the constraints of $(\mathcal{A}(x))$ (cf. (4.27)).

We are not able to analyse the situation in the case when $I_s(x) \neq \emptyset$. However due to numerical tests this technique seems to work well even in this situation.

4.3.4. Numerical Results

In this subsection the aim is to compare the direct nonsmooth approach with the regularization approach introduced in **Haslinger and Neittaanmäki** (1988). In that book the Problems $(\mathcal{P}_l)$ for $l = 1, 2, 3$ were regularized by utilizing the method of penalization (exterior penalty technique) and it was proved that the corresponding optimal designs associated with the penalized problems are close (in an appropriate sense) to optimal designs for the original problems.

The penalized variants of the problems $(\mathcal{P}_l)$ for $l = 1, 2, 3$ read

$$(\mathcal{P}_{l,\varepsilon}) \qquad\qquad \underset{x \in \mathcal{U}}{\text{minimize }} \mathcal{J}_l(x, y_\varepsilon(x))$$

subject to $y_\varepsilon(x)$ solving the penalized state problem

$$(\mathcal{S}_\varepsilon(x)) \qquad\qquad A(x)y_\varepsilon(x) + \frac{1}{\varepsilon} D(y_\varepsilon(x)) = G(x),$$

where $\varepsilon > 0$ is a penalty parameter and the nonlinear operator takes the form

$$D(y_\varepsilon(x)) = -\frac{h}{\varepsilon} \left(0, \ldots, 0, (y^-_{m-n+1,\varepsilon}(x))^2, \ldots, (y^-_{m,\varepsilon}(x))^2 \right).$$

The design sensitivity analysis to Problems $(\mathcal{P}_{l,\varepsilon})$ can be carried out in a standard way applying the adjoint state technique.

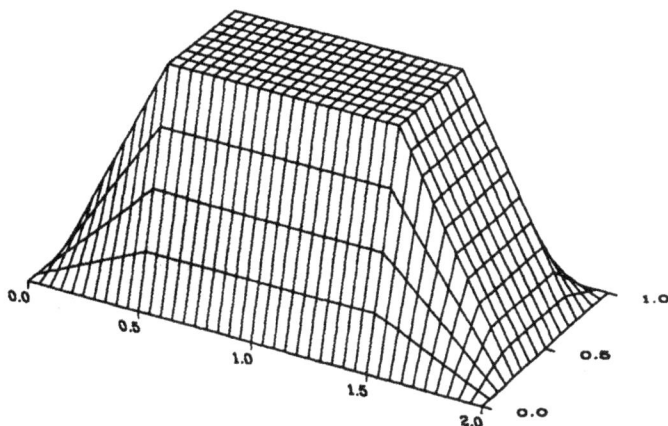

Figure 4.11.

In the following numerical results we have applied the proximal bundle code PB for problems $(\mathcal{P}_l)$ and the subroutine E04VCF (SQP -method) from NAG-library for problems $(\mathcal{P}_{l,\epsilon})$. The state problem $(\mathcal{S}(x))$ has been solved by a projected modification of SOR (successive overrelaxation) -method and $(\mathcal{S}_\epsilon(x))$ by a nonlinear variant of SOR (see **Haslinger and Neittaanmäki** (1988), Appendix I). The test runs have been performed in VAX 4000 computer with double precision relative accuracy $\approx 10^{-17}$.

The discretization parameter h was chosen to be $1/8$, which implies that the corresponding triangulation contains 128 elements. Consequently, the dimensions of the state problem $(\mathcal{S}(x))$ or $(\mathcal{S}_2(x))$ were $n = 9$ and $m = 56$, respectively. The constraint parameters in the set of admissible controls $\mathcal{U}$ were $\alpha = 0.6$, $\beta = 1.5$, $\gamma = M = 1.0$ and the penalty parameter for the regularized problems was $\varepsilon = 10^{-6}$. Moreover, $\alpha' = 0.5$. The starting point was chosen to be $x_i^1 = 1$ for $i = 1, \ldots, n$ i.e. $\Omega(x^1) = (0,1) \times (0,1)$. The function $\Phi \in M(\hat{\Omega})$ in problem $\mathcal{P}_3$ was the same as in **Haslinger, Neittaanmäki** (1988) (see Figure 4.11). As right hand side function g we used

$$g(s_1, s_2) = 8 \sin 4\pi s_1 \sin 4\pi s_2. \tag{4.34}$$

In Figure 4.12 (a) we see the applied coordinate system and in Figure 4.12 (b) the triangulation of the initial design $\Omega(x^1)$, as well as the level sets of spline smoothened FE-solution of the state problem $(\mathcal{S}(x))$ with the right hand side g defined by (4.34).

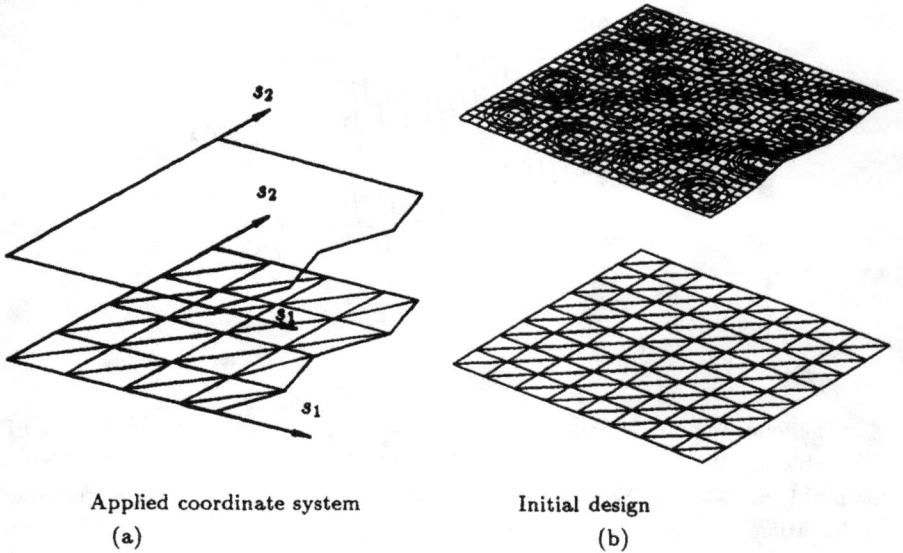

| Applied coordinate system | Initial design |
| (a) | (b) |

Figure 4.12.

Figure 4.13 contains the decrease of the cost functions $(\mathcal{J}_1)$ and $(\mathcal{J}_{1,\epsilon})$ versus the serious steps (see the Proximal Bundle algorithm in part II) obtained by PB and E04VCF. Figures 4.14 (a) and (b) show the triangulation of "optimal" design $\Omega(x)$ for the problems $(\mathcal{P}_1)$ and $(\mathcal{P}_{1,\epsilon})$, respectively, and the level sets of spline smoothened solutions of the state problems.

Our results are summarized in Table 4.1. The following abbreviations will be used: **Initial** = initial value of the cost function, **Final** = final value of the cost function, **it** = number of iterations and **nf** = number of function and subgradient (gradient) evaluations.

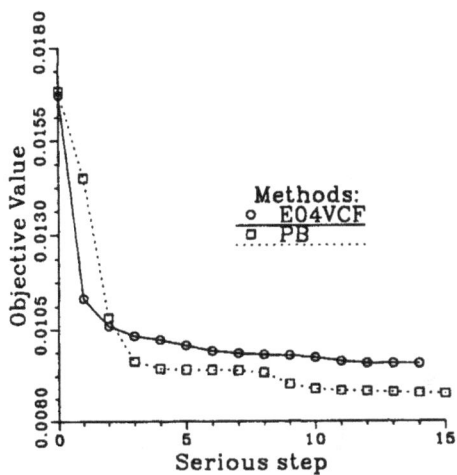

Figure 4.13.

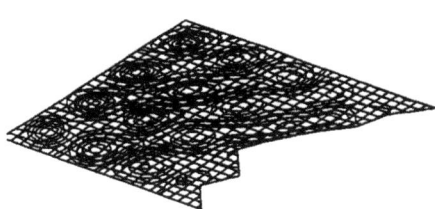

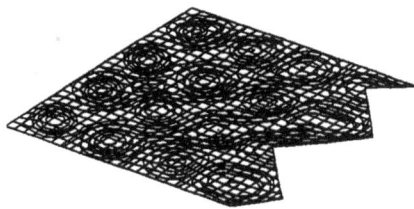

Numerical results for problem $(\mathcal{P}_1)$

(a)

Numerical results for problem $(\mathcal{P}_{1,\epsilon})$

(b)

Figure 4.14.

Table 4.1.

Cost	Initial	Final	it	nf	Algorithm
$\mathcal{J}_1$	$1.68343 \cdot 10^{-2}$	$8.77322 \cdot 10^{-3}$	30	31	PB
$\mathcal{J}_{1,\epsilon}$	$1.67320 \cdot 10^{-2}$	$9.58601 \cdot 10^{-3}$	14	19	E04VCF

Figures 4.15 and 4.16 show the results obtained for the problems $(\mathcal{P}_2)$ and $(\mathcal{P}_{2,\epsilon})$.

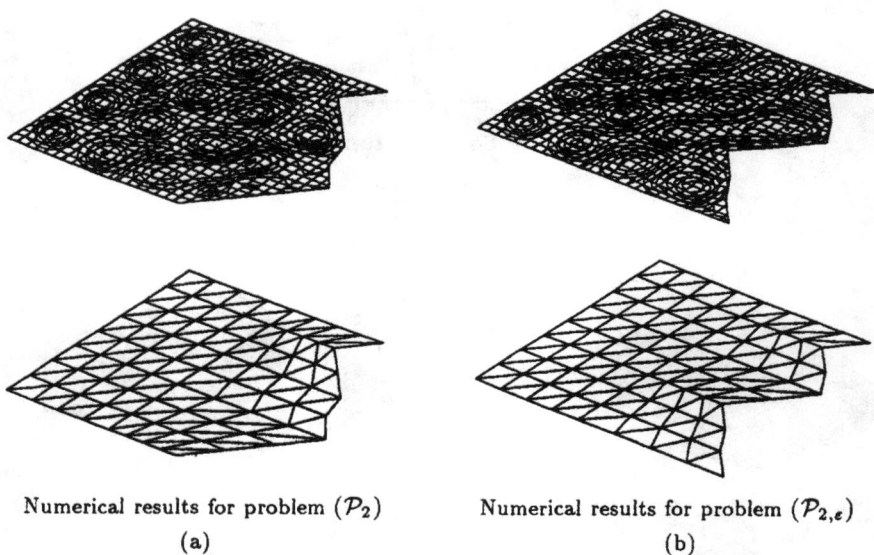

Numerical results for problem $(\mathcal{P}_2)$

(a)

Numerical results for problem $(\mathcal{P}_{2,\epsilon})$

(b)

Figure 4.15.

The results are summarized in Table 4.2.

Table 4.2.

Cost	Initial	Final	it	nf	Algorithm
$\mathcal{J}_2$	$3.92119 \cdot 10^{-2}$	0.0	4	5	PB
$\mathcal{J}_{2,\epsilon}$	$3.91601 \cdot 10^{-2}$	$3.80634 \cdot 10^{-3}$	20	37	E04VCF

Figures 4.17 and 4.18 show the results obtained for the problems $(\mathcal{P}_3)$ and $(\mathcal{P}_{3,\epsilon})$.

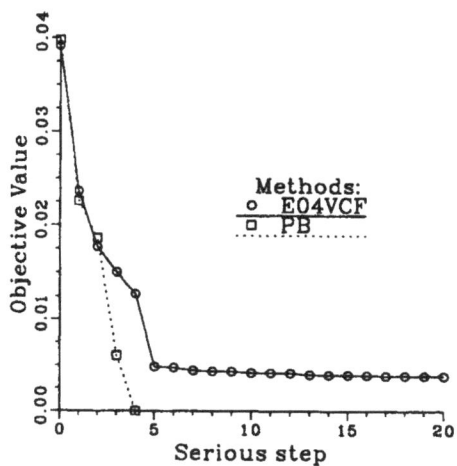

Figure 4.16.

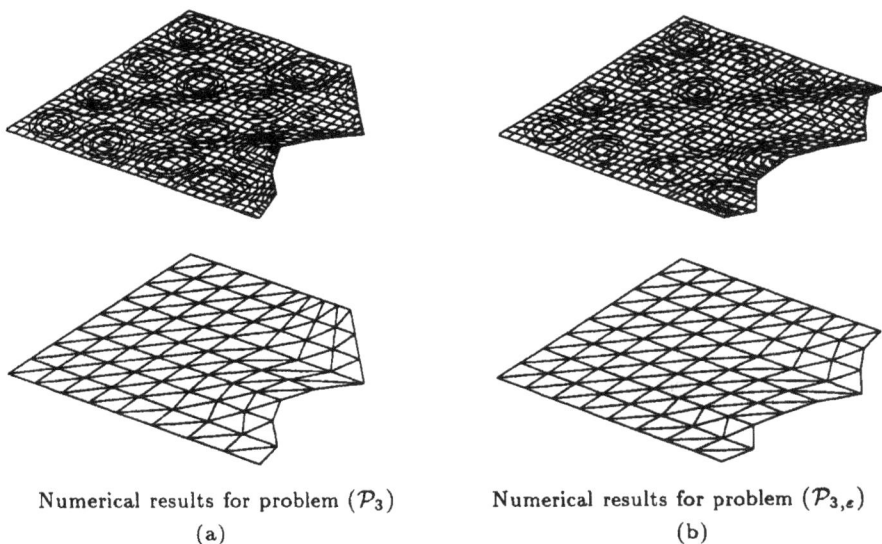

Numerical results for problem $(\mathcal{P}_3)$ Numerical results for problem $(\mathcal{P}_{3,\varepsilon})$

(a)   (b)

Figure 4.17.

The results are again summarized in Table 4.3.

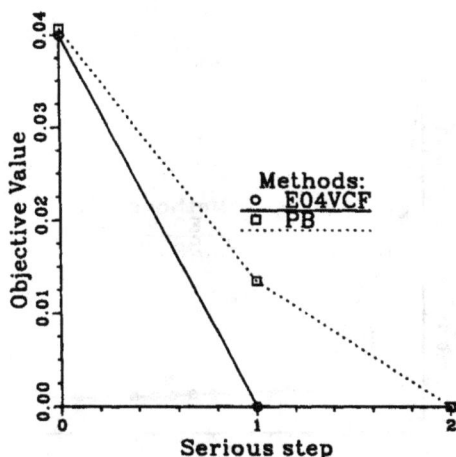

Figure 4.18.

Table 4.3.

Cost	Initial	Final	it	nf	Algorithm
$\mathcal{J}_3$	$4.06385 \cdot 10^{-2}$	$2.07906 \cdot 10^{-8}$	4	5	PB
$\mathcal{J}_{3,\epsilon}$	$3.99266 \cdot 10^{-2}$	$7.16804 \cdot 10^{-8}$	20	37	E04VCF

Remark 4.5. We observed by means of comparable stopping criteria, that the nonsmooth optimization approach give better results. This is caused by the fact that in regularization techniques we always make errors comparable with the penalty parameter ε, thus we actually do not solve the original nonsmooth problem. Furthermore, the solution point typically is located in a nonsmooth corner, which is not attainable by a smoothing process. Although the difference between the final objective values is not very large, the solutions seem to give in some cases slightly different final designs.

4.4. Design of Optimal Covering

In the previous section the state system was described by an unilateral boundary value problem, in which constraints are imposed on the boundary only. Now we shall present a control problem arising from a *free boundary value* problem (obstacle problem).

We consider the problem of controlling the shape of a coincidence set in connection with an obstacle problem. This so-called packaging problem was introduced in **Benedict, Sokolowski and Zolésio** (1984). For further development (existence, finite element approach, penalty method) see **Haslinger and Neittaanmäki** (1988) and we cite also **Neittaanmäki, Tiba and Mäkinen** (1988) (variational inequality approach).

4.4.1. *Setting of the Problem*

Consider a membrane Ω in possible contact with a rigid obstacle Q (see Figure 4.19).

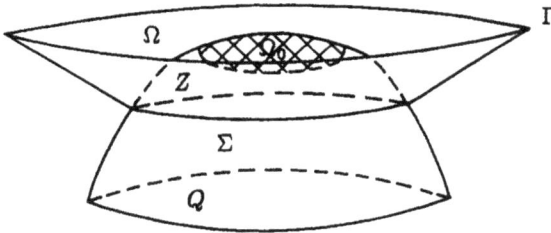

Figure 4.19.

Let us suppose that $\Omega(u)$ is given as in Section 4.2, i.e.

$$\Omega(u) = \{(s_1, s_2) \in \mathbf{R}^2 \mid 0 < s_1 < u(s_2), \ s_2 \in (0, 1)\},$$

where $u \in U_{\text{ad}}$ is a function describing the moving part $\Gamma(u)$ of the boundary Γ, i.e.

$$\Gamma(u) = \{(s_1, s_2) \in \mathbf{R}^2 \mid s_1 = u(s_2), \ s_2 \in (0, 1)\}$$

and

$$U_{\text{ad}} = \{u \in C^{0,1}([0, 1]) \mid 0 < \alpha \le u(s_2) \le \beta, |u'(s_2)| \le \gamma \quad \text{for a.a. } s_2 \in [0, 1]\},$$

where α, β and γ are positive constants chosen in such a way that $U_{\text{ad}} \ne \emptyset$.

Let Q be the graph of a function $q : \mathbf{R}^2 \to \mathbf{R}$. The behaviour of the system is described by

$$-\Delta y(u) \geq g \quad \text{in } \Omega(u) \tag{4.35}$$

$$y(u) \geq q \quad \text{in } \Omega(u) \tag{4.36}$$

$$(-\Delta y(u) - g)(y(u) - q) = 0 \quad \text{in } \Omega(u) \tag{4.37}$$

$$y(u) = 0 \quad \text{on } \Gamma, \tag{4.38}$$

where $g \in L^2(\hat{\Omega})$, $\hat{\Omega} = (0,1) \times (0,\beta)$ and $u \in U_{\text{ad}}$. We suppose that $q \in H^1(\hat{\Omega})$ is a given function such that $q \leq 0$ on $\partial\hat{\Omega}$ and in $(\alpha,\beta) \times (0,1)$. Here $y(u)$ is the vertical displacement of the membrane and g is a given vertical force.

Inequality (4.36) is the nonpenetration condition. For any solution of the problem (4.35)–(4.38) we may define the contact region $Z(y(u))$,

$$Z(y(u)) = \{s \in \Omega(u) \mid y(u;s) = q(s)\},$$

where the contact takes place. Since $\Sigma(y(u))$, the boundary of this subdomain $Z(y(u))$, is not known before the contact problem is solved, it is called a *free boundary*. It is natural to suppose that $\Gamma \cap \Sigma(y(u)) = \emptyset$ (see Figure 4.20). We shall assume that $q \in C(\hat{\Omega})$, too.

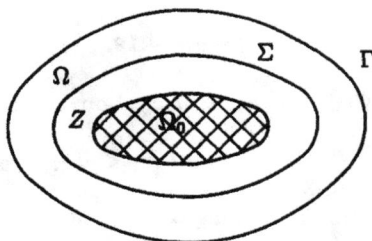

Figure 4.20.

Remark 4.6. By (4.35)–(4.38) we model the deflection of a membrane that may, under a vertical load g, come into contact with a rigid obstacle described by the function q.

In what follows we shall analyse the numerical realization of the *packaging problem* — the design problem of minimizing the area of $\Omega(u)$, $u \in U_{ad}$, such that the contact region $Z(y(u))$ of the corresponding solution $y(u)$ contains the specific region Ω_0. Let us denote

$$\tilde{U}_{ad} = \{u \in U_{ad} \mid Z(y(u)) \supseteq \Omega_0\}$$

and

$$J(u) = \text{meas } \Omega(u),$$

where meas $\Omega(u)$ means the measure of the domain $\Omega(u)$. The mathematical formulation of the optimal covering problem reads as follows

(**P**)
$$\underset{u \in \tilde{U}_{ad}}{\text{minimize}} J(u)$$

subject to $y(u) \in K = K(u)$ solving the variational inequality

(**S**(u))
$$a_u(y(u), v - y(u)) \geq (g, v - y(u))_{0,\Omega(u)} \quad \text{for all} \quad v \in K,$$

where

$$K = \{v \in H_0^1(\Omega(u)) \mid v \geq q \text{ a.e. in } \Omega(u)\}$$

and

$$a_u(y, v) = \int_{\Omega(u)} \nabla y \nabla v \, d\Omega(u)$$

$$(g, v)_{0,\Omega(u)} = \int_{\Omega(u)} g v \, d\Omega(u).$$

Using Green's formula on (**S**(u)), we formally obtain the relations (4.35)–(4.38) in $\Omega(u)$. Concerning the solvability of (**P**) we have the following result.

Theorem 4.6. *Let* $\tilde{U}_{ad} \neq \emptyset$. *Then there exists at least one solution of* (**P**).

Proof. See **Haslinger and Neittaanmäki** (1988), Chapter 9. □

The main difficulty in the numerical realization of (**P**) is the presence of the state constraint $y(u) = q$ in Ω_0. One way to overcome this difficulty is to use the penalty approach with exact penalty function.

Let

$$J_\varepsilon(u, y(u)) = \text{meas } \Omega(u) + \frac{1}{\varepsilon} \int\limits_{\Omega_0} (y(u) - q) \, d\Omega_0, \quad \varepsilon > 0,$$

be a modified cost functional with a penalty term $\frac{1}{\varepsilon} \int_{\Omega_0} (y(u) - q) \, d\Omega_0$ where $y(u)$ is the solution of $(\mathbf{S}(u))$.

The penalty form of $(\mathbf{P})$ now reads as follows

$(\mathbf{P_\varepsilon})$ $\underset{u \in U_{\text{ad}}}{\text{minimize }} J_\varepsilon(u, y(u)),$

where $y(u) \in K$ is the solution of the state problem $(\mathbf{S}(u))$.

The relation between solutions of $(\mathbf{P})$ and $(\mathbf{P_\varepsilon})$ for $\varepsilon \to 0$ is studied in **Haslinger and Neittaanmäki** (1988), Chapter 9.

4.4.2. Discretization of the Problem

As in Section 4.3 we shall use linear finite elements for approximating y and u. The discretization can be done as in the previous example (subsection 4.3.2). Again let $0 = a_1 < a_2 < \cdots < a_n = 1$ be an equidistant partition of $[0, 1]$, thus $a_i = (i - 1)h$ for $i = 1, \ldots, n$. The discretization of U_{ad} is defined as follows

$$U_{\text{ad}}^h = \{u_h \in U_{\text{ad}} \mid u_h|_{[a_{i-1}, a_i]} \in P_1, \; i = 2, \ldots, n\},$$

i.e. U_{ad}^h contains all functions from U_{ad} which are piecewise linear on a given partition of $[0, 1]$. By $T(h, u_h)$ we denote a uniformly regular triangulation of $\overline{\Omega(u_h)}$, i.e. $\overline{\Omega(u_h)} = \cup_{T \in T(h, u_h)} T$, $u_h \in U_{\text{ad}}^h$. Moreover, we suppose as in subsection 4.3.2 that $T(h, u_h)$ depends continuously on u_h. In what follows we assume that Ω_0 is a *polygonal domain*, covered by a finite number of $T \in T(h, u_h)$ and it is located in the area of fixed nodes i.e. $\Omega_0 \subset \Omega(\alpha') = (0, \alpha') \times (0, 1)$, where $\alpha' < \alpha$ (for a more general situation we refer to **Haslinger and Neittaanmäki** (1988)). We denote the nodes of $\Omega(u_h)$ by N_i for $i \in I$, where I is the set of indices. We split I in this case as follows

$$I = I_{\Omega(u_h)} = I_{\overline{\Omega}_0} \cup I_{\Omega(u_h) \setminus \overline{\Omega}_0}.$$

Let

$$m = m(h) = \text{ card } (I_{\Omega(u_h)}).$$

Thus, in this case, m is the number of the nodes lying on $\Omega(u_h)$ and n is, again, the number of the design nodes N_i^D lying on $\overline{\Gamma(u_h)}$. Moreover, $I_{\overline{\Omega}_0}$ is the set of indices associated with the nodes lying in the polygonal domain $\overline{\Omega}_0$, where contact with the obstacle is required.

The discretized variant of $(\mathbf{P}_\varepsilon)$ reads

$$(\mathbf{P}_{\varepsilon,h}) \qquad \underset{u_h \in U_{ad}^h}{\text{minimize}} \left\{ J_{\varepsilon,h}(u_h, y_h) \equiv \text{meas } \Omega(u_h) + \frac{1}{\varepsilon} \int\limits_{\Omega_0} (y_h(u_h) - q) \, d\Omega_0 \right\},$$

subject to $y_h = y_h(u_h) \in K_h$ solving the variational inequality

$$(\mathbf{S}_h(u_h)) \qquad a_{u_h}(y_h(u_h), v_h - y_h(u_h)) \geq (g, v_h - y_h(u_h))_{0,\Omega(u_h)} \quad \text{for all } v_h \in K_h,$$

where

$$K_h = \{v_h \in C(\overline{\Omega(u_h)}) \mid v_h|_T \in P_1 \ \forall \, T \in \mathcal{T}(h, u_h),$$
$$v_h \geq q \text{ in } \Omega(u_h) \text{ and } v_h = 0 \text{ on } \partial\Omega(u_h)\}.$$

It is not difficult to prove that $(\mathbf{P}_{\varepsilon h})$ has at least one solution. The relation between solutions of $(\mathbf{P}_\varepsilon)$ and $(\mathbf{P}_{\varepsilon h})$ for $h \to 0_+$ is studied in **Haslinger and Neittaanmäki** (1988).

In order to present $(\mathbf{P}_{\varepsilon,h})$ in an algebraic form, we define again the design variables $x_i = u_h(a_i)$ for $i = 1, \ldots, n$. Further let $y_i = y_h(N_i)$ and $q_i = q(N_i)$ for $i = 1, \ldots, m$. The sets U_{ad}^h and K_h can be identified with closed convex sets

$$\mathcal{U} = \{x \in \mathbf{R}^n \mid 0 < \alpha \leq x_i \leq \beta, \ i = 1, \ldots, n,$$
$$-\gamma h \leq x_{i+1} - x_i \leq \gamma h, \ i = 1, \ldots, n-1\} \qquad (4.39)$$
$$\mathcal{K} = \{y \in \mathbf{R}^m \mid y_i \geq q_i, \ i = 1, \ldots, m\}.$$

The algebraic form of the state problem $(\mathbf{S}_h(u_h))$ for a fixed $x \in \mathcal{U}$ can be formulated as follows: Find $y = y(x) \in \mathcal{K}$ such that

$$(\mathcal{S}(x)) \qquad (v - y(x))^T A(x)(v - y(x)) \geq G(x)^T (v - y(x)) \quad \text{for all } v \in \mathcal{K}$$

or equivalently: Find $y = y(x) \in \mathcal{K}$ such that

$$(\mathcal{S}(x)') \qquad y = \arg\min_{z \in \mathcal{K}} \left\{ \tfrac{1}{2} z^T A(x) z - G(x)^T z \right\},$$

where

$$A(x) = \left(A_{ij}(x) \right)_{i,j=1}^m = \left(\int\limits_{\Omega(u_h)} \nabla\varphi_i \nabla\varphi_j \, d\Omega(u_h) \right)_{i,j=1}^m$$

is the symmetric, positive definite stiffness matrix and

$$G(x) = \left(G_j(x)\right)_{j=1}^m = \left(\int_{\Omega(u_h)} g\varphi_j \, d\Omega(u_h)\right)_{j=1}^m$$

is the linear term, arising from the discretization of the right hand side g.

The problem $(P_{\varepsilon,h})$ can now be expressed in terms of nodal values by

$$(\mathcal{P}_\varepsilon) \qquad \underset{x \in \mathcal{U}}{\text{minimize}} \left\{ \mathcal{J}_\varepsilon(x, y(x)) \equiv \frac{h}{2}\sum_{i=1}^{n-1}(x_i + x_{i+1}) + \frac{1}{2\varepsilon}\sum_{i \in I_{\Omega_0}}(y_i - q_i) \right\},$$

where $y(x)$ solves the state problem $(\mathcal{S}(x))$.

4.4.3. Design Sensitivity Analysis

For performing the design sensitivity analysis we shall again apply Theorem 2.1. We rewrite $\mathcal{J}_\varepsilon(x, y(x))$ in the form

$$\mathcal{J}_\varepsilon(x, y(x)) = h(x) + E(y(x)) = C^{\mathrm{T}}x + \frac{1}{\varepsilon}B(y(x) - q), \tag{4.40}$$

where

$$C = \frac{h}{2}(1, 2, 2, \dots, 2, 1)^{\mathrm{T}} \in \mathbf{R}^n,$$

$$B = (b_1, \dots, b_n)^{\mathrm{T}} \in \mathbf{R}^m, b_i = \begin{cases} 1, & \text{if } i \in I_{\overline{\Omega}_0}, \\ 0, & \text{if } i \notin I_{\overline{\Omega}_0}. \end{cases}$$

According to Theorem 2.1

$$C + \frac{1}{\varepsilon}\xi_y(x)^{\mathrm{T}}B \in \partial \mathcal{J}_\varepsilon(x, y(x)). \tag{4.41}$$

As before Karush–Kuhn–Tucker optimality conditions, applied to $(\mathcal{S}(x)')$, guarantee the existence of a non-negative Lagrange multiplier $\lambda = \lambda(x) \in \mathbf{R}^m$ such that

$$A(x)y(x) = G(x) + \lambda(x), \quad \lambda_i(x) \geq 0, \ i \in I_{\Omega(u_h)}$$

$$(y_i(x) - q_i)\lambda_i(x) = 0 \qquad \text{for all} \quad i \in I_{\Omega(u_h)}. \tag{4.42}$$

We divide now $I = I_{\Omega(u_h)}$ into three parts

$$I_a(x) = \{i \in I \mid y_i(x) = q_i \text{ and } \lambda_i(x) > 0\}$$

$$I_n(x) = \{i \in I \mid y_i(x) > q_i \text{ and } \lambda_i(x) = 0\}$$

$$I_s(x) = \{i \in I \mid y_i(x) = q_i \text{ and } \lambda_i(x) = 0\}.$$

The next result follows from Appendix II of **Haslinger and Neittaanmäki** (1988).

Theorem 4.7. *Let $y(x) \in \mathcal{K}$ be the solution of the state problem $(S(x)')$. Then the directional derivative $y'(x; d) = (y'_1(x; d), \ldots, y'_m(x; d))^T$ at $x \in \mathcal{U}$ in the direction $d \in \mathbf{R}^n$ exists and can be found as the unique solution of the quadratic programming problem*

$$\underset{z \in \tilde{\mathcal{K}}(x)}{\text{minimize}} \left\{ \tfrac{1}{2} z^T A(x) z - z^T (\nabla G(x) d - \nabla A(x)[y(x)] d) \right\}, \tag{4.43}$$

where

$$\tilde{\mathcal{K}}(x) = \{ z \in \mathbf{R}^m \mid z_i = 0 \quad \text{for all} \quad i \in I_a(x)$$
$$z_i \geq 0 \quad \text{for all} \quad i \in I_s(x) \}.$$

We notice, that $y'(x; d)$ is nonlinear with respect to d, if $I_s(x) \neq \emptyset$ and therefore y is only directionally differentiable at x in this case.

Now we introduce the adjoint state $p(x) \in \mathbf{R}^m$ (when exists) to Problem $(\mathcal{P})$ as the solution of the quadratic programming problem

$$(\mathcal{A}(x)) \qquad \underset{p \in \tilde{\mathcal{K}}(x)}{\text{minimize}} \left\{ \tfrac{1}{2} p^T A(x) p - \frac{1}{\varepsilon} p^T B \right\}.$$

Application of the Karush–Kuhn–Tucker optimality conditions to $(\mathcal{A}(x))$ yields a vector of Lagrange multipliers $\mu(x) \in \mathbf{R}^m$ such that

$$A(x) p(x) = \frac{1}{\varepsilon} B + \mu(x)$$
$$p_i(x) = 0 \quad \text{for all} \quad i \in I_a(x)$$
$$\mu_i(x) = 0 \quad \text{for all} \quad i \in I_n(x) \tag{4.44}$$
$$\mu_i(x) \geq 0, p_i(x) \geq 0, p_i(x)\mu_i(x) = 0 \quad \text{for all} \quad i \in I_s(x).$$

By means of (4.41) and (4.44) we obtain

$$C + \frac{1}{\varepsilon} \xi_y(x)^T (A(x) p(x) - \mu(x))$$
$$= C + \frac{1}{\varepsilon} (A(x) \xi_y(x))^T p(x) - \xi_y(x)^T \mu(x) \in \partial \mathcal{J}(x, y(x)). \tag{4.45}$$

Due to (4.42) we have

$$\partial(A(x) y(x)) = \partial G(x) + \partial \lambda(x), \tag{4.46}$$

implying

$$A(x)\xi_y(x) = \nabla G(x) - \nabla A(x)[y(x)] + \xi_\lambda(x)$$

for some $\xi_\lambda(x) \in \partial\lambda(x)$. Consequently,

$$C + \frac{1}{\varepsilon}(\nabla G(x) - \nabla A(x)[y(x)] + \xi_\lambda(x))^T p(x) - \xi_y(x)^T \mu(x) \in \partial \mathcal{J}(x, y(x)). \quad (4.47)$$

As in the previous section we conclude that $\xi_\lambda(x)^T p(x) = 0$ and $\xi_y(x)^T \mu(x) = 0$, if $i \in I_a(x) \cup I_n(x)$. Thus we have

$$\nabla \mathcal{J}(x, y(x)) = C + \frac{1}{\varepsilon}(\nabla G(x) - \nabla A(x)[y(x)])^T p(x), \quad (4.48)$$

if $I_s(x) = \emptyset$ (the differentiable case). Also in this problem we are not able to analyse further the situation in the case, when $I_s(x) \neq \emptyset$. However, due to numerical tests, this technique seems to work well (as well in the case $I_s(x) \neq \emptyset$).

4.4.4. Numerical Results

We have applied similar triangulation as in Subsection 4.3.4 with $h = 1/8$. This triangulation contains 128 elements and the dimensions of the state problem and the optimal shape design problem are $m = 56$ and $n = 9$, respectively. The constraint parameters in the set of admissible controls $\mathcal{U}$ are $\alpha = 0.6$, $\beta = 1.5$ and $\gamma = 1.0$.

The state problem $(\mathcal{S}(x))$ and the adjoint state problem $(\mathcal{A}(x))$ have been solved by using the code MEMBRANE by Outrata. In this code the state problem $(\mathcal{S}(x))$ and the adjoint state problem $(\mathcal{A}(x))$ has been solved by SOR with projection. To solve the optimization problem $(\mathcal{P})$ we have used the code PB with the penalty parameter $\varepsilon = 10^{-4}$. The test runs have been performed in VAX 4000 computer with double precision relative accuracy $\approx 10^{-17}$.

Example 4.3. Let the obstacle be defined by the function

$$q(s_1, s_2) = -\frac{1}{20}(s_1^2 + (s_2 - \frac{1}{4})^2).$$

The contour plot of the obstacle is illustrated in Figure 4.21.

Let the desired contact region to be the rectangle

$$\Omega_0 = \left(\frac{1}{4}, \frac{1}{2}\right) \times \left(\frac{1}{4}, \frac{3}{4}\right)$$

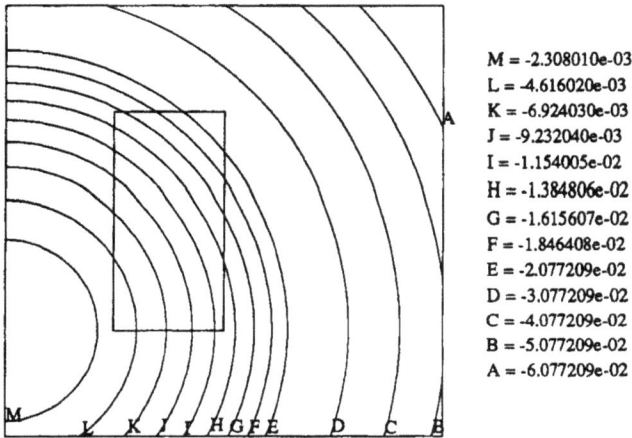

$M = -2.308010e-03$
$L = -4.616020e-03$
$K = -6.924030e-03$
$J = -9.232040e-03$
$I = -1.154005e-02$
$H = -1.384806e-02$
$G = -1.615607e-02$
$F = -1.846408e-02$
$E = -2.077209e-02$
$D = -3.077209e-02$
$C = -4.077209e-02$
$B = -5.077209e-02$
$A = -6.077209e-02$

Figure 4.21.

and let $g \equiv -1$. We have chosen the initial guess to be the unit square, i.e. $x_i^1 = 1$ for $i = 1, \ldots, 9$. In Figure 4.22 (a) we see the triangulation of $\Omega(x^1)$ and in Figure 4.22 (b) the contour plot of the corresponding state (Ω_0 is indicated by quadrangle).

Figure 4.23 (a) shows the triangulation of obtained optimal domain $\Omega(x^{45})$ and Figure 4.23 (b) the contour plot of corresponding state. By comparing Figure 4.21 and 4.23 (b) we find that equipotential contours of the state and the coincides in Ω_0. The results can be compared with those obtained in **Haslinger and Neittaanmäki** (1988).

Example 4.4. Let the obstacle be given by the function

$$q(s_1, s_2) = -0.03 \sin(\pi s_1) \sin(\pi s_2).$$

Figure 4.24 shows the contour plot of the obstacle.

Let the rest of the data be the same as in the previous example. We shall also use similar mesh. We have chosen the initial guess $x_i^1 = 0.7$ for $i = 1, \ldots, 9$. Figure 4.25 shows the triangulation of $\Omega(x^1)$ as well as the contour plot of corresponding state.

By comparing Figure 4.24 and 4.25 (b) we find that the state constraint is not satisfied.

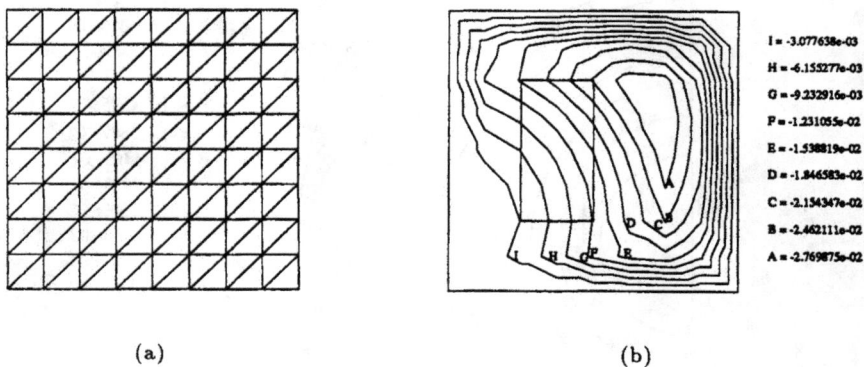

(a)	(b)

Figure 4.22.

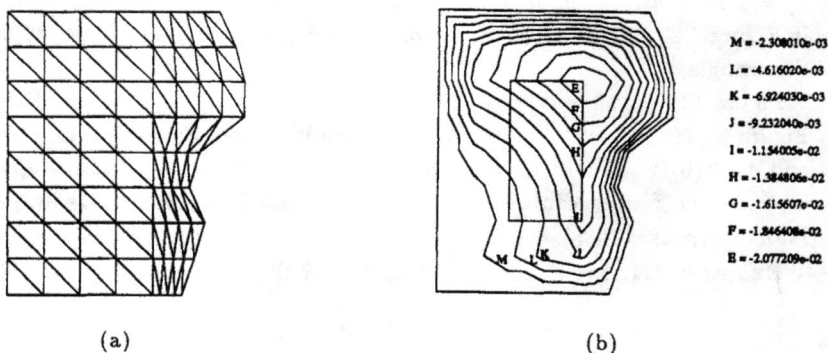

(a)	(b)

Figure 4.23.

Figure 4.26 shows the triangulation of the obtained optimal domain $\Omega(x^{27})$ and the contour plot of the corresponding state. The results can be compared with those obtained in **Neittaanmäki, Tiba and Mäkinen** (1988).

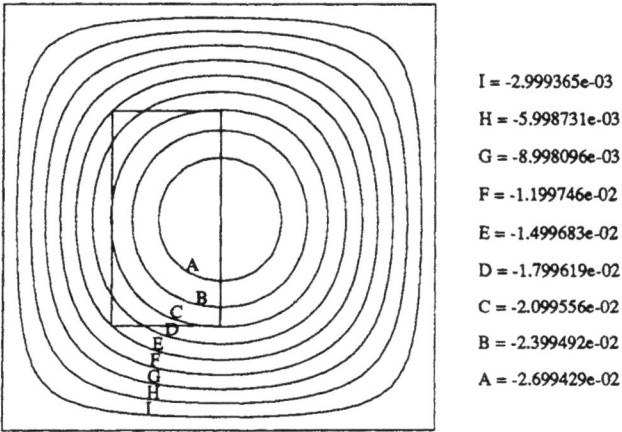

I = -2.999365e-03

H = -5.998731e-03

G = -8.998096e-03

F = -1.199746e-02

E = -1.499683e-02

D = -1.799619e-02

C = -2.099556e-02

B = -2.399492e-02

A = -2.699429e-02

Figure 4.24.

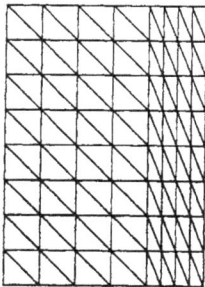

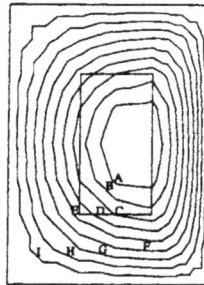

I = -2.771638e-03
H = -5.543277e-03
G = -8.314915e-03
F = -1.108655e-02
E = -1.385819e-02
D = -1.662983e-02
C = -1.940147e-02
B = -2.217311e-02
A = -2.494475e-02

(a)   (b)

Figure 4.25.

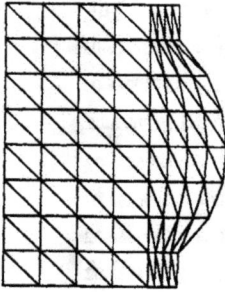

(a)   (b)

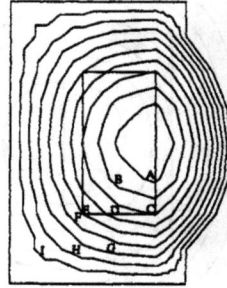

I = -2.999365e-03
H = -5.998731e-03
G = -8.998096e-03
F = -1.199746e-02
E = -1.499683e-02
D = -1.799619e-02
C = -2.099556e-02
B = -2.399492e-02
A = -2.699429e-02

Figure 4.26.

Chapter 5

Boundary Control for Stefan Type Problems

5.1. Introduction

Optimal control problems associated with nonlinear parabolic boundary value problems are of great importance in a variety of technological applications. In particular such problems are arising in mechanics, heat transfer and the theory of free boundary problems. Examples of this kind of problems are heat conduction, heat radiation, natural convection, enzyme diffusion, the thermostat-control process and the Signorini problem.

Results on existence, necessary optimality conditions and approximation for control problems governed by nonlinear parabolic equations have been presented in **Tiba** (1990) and **Tröltzsch** (1984).

5.2. Setting of the Abstract Problem

We shall begin with a simple boundary control problem of Stefan type. Consider the boundary control problem

$$(\mathbf{P}) \qquad\qquad \underset{u \in U_{\mathrm{ad}}}{\operatorname{minimize}} J(u, y(u)),$$

where

$$J(u, y) = \int\limits_{0}^{T} \left\{ \tfrac{1}{2} \|y - y_d\|_{0,\Omega}^2 + \tfrac{1}{2}\|u\|_{0,\Gamma}^2 \right\} dt \qquad (5.1)$$

subject to

$$\frac{\partial}{\partial t} v(t,s) - \Delta y(t,s) = g(t,s) \qquad \text{in } Q \qquad (5.2)$$

$$v(t,s) \in H(y(t,s)) \qquad \text{in } Q$$

$$\frac{\partial}{\partial n} y(t,s) = u(t,s) \qquad \text{on } \Sigma \qquad (5.3)$$

$$y(0,s) = v_0(s) \qquad \text{in } \Omega. \qquad (5.4)$$

In this setting, $\Omega \subset \mathbf{R}^n$, $n \geq 1$, is a bounded domain with boundary Γ and $Q = (0,T) \times \Omega$ is a cylinder with lateral face Σ. We assume that $y_d, g \in L^2(Q)$,

$v^0 \in L^2(\Omega)$ and that H is a strongly maximal monotone graph in $\mathbf{R} \times \mathbf{R}$, i.e. there exists $\alpha > 0$ such that

$$(H(y) - H(z))(y - z) \geq \alpha \|y - z\|^2, \tag{5.5}$$

and H is bounded on bounded sets.

When H is given by

$$H(y) = \begin{cases} y - r_0; & y > r_0 \\ [-\sigma, 0]; & y = r_0 \\ \kappa(y - r_0) - \sigma; & y < r_0, \end{cases} \tag{5.6}$$

where $\kappa, \sigma > 0$, then y solves the two-phase Stefan problem (5.2)–(5.4) for given boundary control $u \in U_{ad} = L^2(\Sigma)$ (no constraints). The physical meaning of the above variables is the following:

u – boundary heat flux (control variable)

y – temperature (state variable)

v – enthalpy

g – internal heat source.

Example 5.1. Let $\Omega = (0,1) \times (0,1)$, $T = 1$ and $U_{ad} = L^2(\Gamma)$. Moreover, we suppose that

$$H(y) = \begin{cases} y; & y < 0 \\ [0,2]; & y = 0 \\ 4y + 2; & y > 0 \end{cases} \tag{5.7}$$

$$g(t,s) = \begin{cases} 8(2e^{-2t} - 1); & s_1^2 + s_2^2 \geq e^{-2t} \\ 2(2e^{-2t} - 2); & s_1^2 + s_2^2 \leq e^{-2t} \end{cases} \tag{5.8}$$

$$v_0 = H(y_0)$$

$$y_0(s) = \begin{cases} s_1^2 + s_2^2 - 1; & s_1^2 + s_2^2 < 1 \\ 2(s_1^2 + s_2^2 - 1); & s_1^2 + s_2^2 \geq 1 \end{cases}$$

$$u(t,s) = \begin{cases} 0; & \text{on the axes} \\ 4; & \text{on the parallels to the axes.} \end{cases} \tag{5.9}$$

As from (5.7) follows that the phase change temperature is 0, we take $y_d = 0$. The classical solution of the Stefan free boundary problem (5.2)–(5.4) with above data (5.7)–(5.9) is

$$y(t,s) = \begin{cases} 2(s_1^2 + s_2^2 - e^{-2t}) & \text{if } s_1^2 + s_2^2 \geq e^{-2t}, \\ s_1^2 + s_2^2 - e^{-2t} & \text{if } s_1^2 + s_2^2 < e^{-2t}. \end{cases}$$

In Figure 5.1 we see a lateral face $\Sigma = \{(t, s_1, s_2) \in \mathbf{R}^3 \mid t \in (0,1), s_1 \in (0,1), s_2 = 0\}$. The free boundary $\{(t, s_1, s_2) \in \Sigma \mid y(t, s_1, s_2) = 0\}$ can be seen in the figure as well.

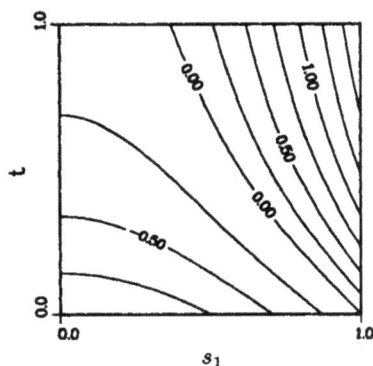

Figure 5.1.

5.3. Simulation of the Continuous Casting Process

We introduce a mathematical model which is used to simulate the continuous casting process and to control the secondary cooling water spray nozzles. The main object is to minimize the defects in the final products. The problem is formulated as an optimal control problem where the cost function is constructed according to certain metallurgical criteria and constraints. The temperature distribution of the strand is calculated by solving a nonlinear heat transfer equation with free boundaries between solid and liquid phases.

The state constraints are penalized by using exact penalty functions, which are nonsmooth. Due to the piecewise linear approximation of nonlinear terms, the cost function is nonsmooth even without penalties.

We shall now shortly describe the physical situation, for more detailed presentation see **Neittaanmäki and Laitinen** (1988).

5.3.1. Setting of the Problem

The control variable u represents a heat transfer coefficient, which has an effect on the temperature distribution $y(u)$ (the state) of the steel strand. Our aim

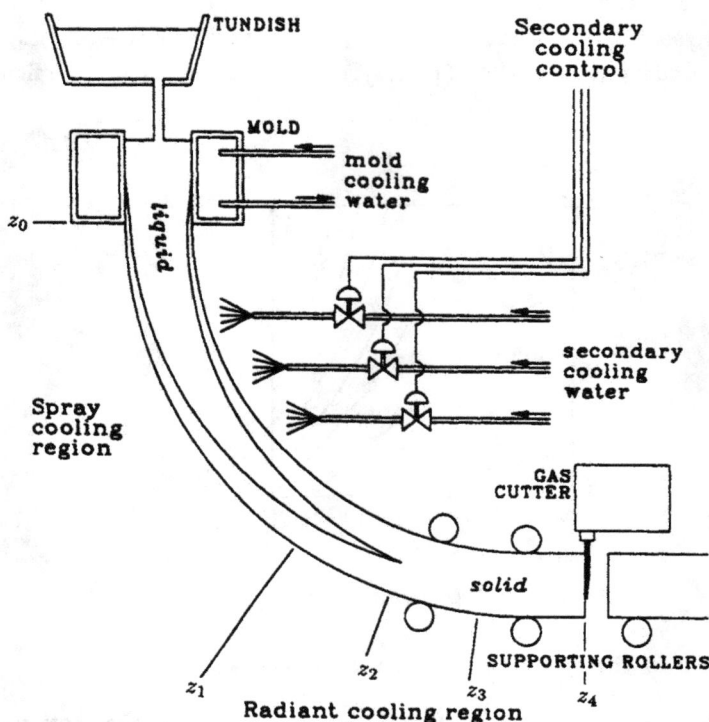

Figure 5.2. Schematic representation of the continuous casting process.

is to maximize the quality of the final product. This is realized by keeping the temperature $y(u)$ as close as possible to some desired temperature distribution y_d.

The essential features of a continuous casting machine are shown in Figure 5.2. Molten steel is poured down from the tundish into the water cooled mold, where the metal gets a solid shell. After the end of the mold (point z_0) the strand is supported by rollers and cooled down by water sprays so that eventually the solidification is completed (the maximum length of the liquid pool is denoted by z_2). After the water sprays (point z_1) the strand is cooled down only by radiation. The strand is straightened at the unbending point z_3 and in the cutting point z_4 it is cut up.

Let $\Omega \subset \mathbf{R}^2$ denote the cross-section of the strand and $\Gamma = \partial\Omega$ its boundary. We define t_i for $i = 0, \ldots, 3$ to be the time events when Ω passes the distances z_i (cf. the Figure 5.2) and we consider the time period between t_0 and t_3, i.e. $t_0 = 0$ and $t_3 = T$. Moreover, we denote $Q = (0, T) \times \Omega$, $\Sigma = (0, T) \times \Gamma$, $Q_1 = (t_2, T) \times \Omega$

and $\Sigma_1 = (0, t_1) \times \Gamma$. The set of admissible controls is defined by

$$U_{ad} = \{u \in L^2(\Sigma) \mid 0 < \alpha(t) \leq u(t, s) \leq \beta(t), \ t \in (0, T), \ s \in \Gamma\}.$$

Then for $u \in U_{ad}$ the temperature distribution $y = y(u)$ is obtained by solving the state system

$$\text{(S}(u)\text{)} \qquad \begin{aligned} &\frac{\partial}{\partial t} H(y(u)) - \Delta K(y(u)) = 0 &&\text{in } Q \\ &\frac{\partial}{\partial n} K(y(u)) = g(u, y(u)) &&\text{on } \Sigma \\ &y(s, 0; u) = y_0(s) &&s \in \Omega, \end{aligned}$$

where

$$g(u, y(u)) = \begin{cases} u \cdot (y_{wat} - y(u)) + c \cdot (y_{ext}^4 - y(u)^4) & \text{on } \Sigma_1 \\ c \cdot (y_{ext}^4 - y(u)^4) & \text{on } \Sigma \setminus \Sigma_1. \end{cases}$$

We suppose that the enthalpy function H and the Kirchhoff's transformation K are piecewise linear functions, for more details we refer to **Neittaanmäki and Laitinen** (1988) and Subsection 5.3.5. Let the constants y_0, y_{wat} and y_{ext} denote the initial, the spray water and the surrounding environment temperature, respectively, and let c be some physical constant.

We define on the boundary of the strand some temperature distribution $y_d = y_d(t, s)$, which in technological sense is good, and we want the actual surface temperature to be as close to y_d as possible. Thus, our problem now reads

$$\text{(P)} \qquad \underset{u \in U_{ad}}{\text{minimize}} \left\{ J(y(u)) \equiv \int_0^T \tfrac{1}{2} \|y(u) - y_d\|_{0,\Gamma}^2 \, dt \right\}$$

where $y(u)$ solves the state system $(\text{S}(u))$ and moreover, we have the following technological constraints

$$y_{min} \leq y(u) \leq y_{max} \qquad \text{on } \Sigma \qquad (5.10)$$

$$y'_{min} \leq \frac{\partial}{\partial t} y(u) \leq y'_{max} \qquad \text{on } \Sigma \qquad (5.11)$$

$$0 \leq y(u) \leq y_{sol} \qquad \text{in } Q_1 \qquad (5.12)$$

$$y_{duc} \leq y(\cdot, T; u) \leq y_{sol} \qquad \text{in } \Omega. \qquad (5.13)$$

Here the constants y_{min}, y_{max}, y'_{min}, y'_{max}, y_{sol} and y_{duc} denote some minimum and maximum bounds (also for derivatives), the solidus and the ductility temperature respectively.

5.3.2. Penalization of the State Constraints

As before, we are using the exact penalty technique to handle the additional state constraints (5.10)–(5.13). We define the penalty functions

$$\Psi_1(y(u)) = \left[\int_0^T \|y(u) - \Pi_1[y(u)]\|_{0,\Gamma}^2 \, dt \right]^{\frac{1}{2}},$$

$$\Psi_2(y(u)) = \left[\int_0^T \left\| \frac{\partial}{\partial t} y(u) - \Pi_2[\frac{\partial}{\partial t} y(u)] \right\|_{0,\Gamma}^2 \, dt \right]^{\frac{1}{2}}, \qquad (5.14)$$

$$\Psi_3(y(u)) = \left[\int_{t_2}^T \|y(u) - \Pi_3[y(u)]\|_{0,\Omega}^2 \, dt \right]^{\frac{1}{2}},$$

$$\Psi_4(y(u)) = \|y(\cdot, T; u) - \Pi_4[y(\cdot, T; u)]\|_{0,\Omega},$$

where Π_i for $i = 1, \ldots, 4$ denote the projections into the feasible sets of (5.10)–(5.13), respectively.

The penalized version of problem (**P**) now reads

$$(\mathbf{P_\epsilon}) \qquad \operatorname*{minimize}_{u \in U_{ad}} \left\{ J_\epsilon(y(u)) \equiv J(y(u)) + \sum_{i=1}^4 \frac{1}{\epsilon_i} \Psi_i(y(u)) \right\}$$

where $\epsilon_i > 0$ are the penalty parameters and $y(u)$ solves the system (**S**(u)).

5.3.3. Discretization of the Problem

The problem (**P$_\epsilon$**) is discretized by using the finite element method (FEM) with respect to space and the finite difference method (FD) with respect to time. We give only some basic ideas of the discretization process, the more detailed presentation can be found in **Křížek and Neittaanmäki (1990)** and **Neittaanmäki and Laitinen (1988)** .

Let T_h be a regular triangulation of $\overline{\Omega}$, where the discretization parameter h denotes the maximum length of the triangle edges. Let I_Ω denote the index set and N_i for $i \in I_\Omega$ denote the nodes of T_h. The time interval $(0, T)$ will be divided into k_3 subintervals with the points $0 = t^0 < t^1 < \cdots < t^{k_3-1} < t^{k_3} = T$, such that the time events t_i, $i = 1, 2$, are also among those points (i.e., $t_i = t^{k_i}$ with some k_i, $i = 0, \ldots, 3$). Furthermore, we denote $\Delta t^k = t^k - t^{k-1}$, $k = 1, \ldots, k_3$ and

$$m_\Omega = m_\Omega(h) = \text{card } (I_\Omega),$$
$$n_\Gamma = n_\Gamma(h) = \text{card } (I_\Gamma),$$
$$m = m(h) = k_3 \cdot m_\Omega,$$
$$n = n(h) = k_3 \cdot n_\Gamma,$$

thus m_Ω, n_Γ, m and n are the number of nodes lying on $\overline{\Omega}$, Γ, $\overline{Q}$ and Σ respectively. We use the piecewise linear finite element approximation for the state $y(u)$ and control u. Consequently we have the vectors y and x, respectively, such that

$$y = (y^1, \ldots, y^{k_3})^T \in \mathbf{R}^m \quad \text{and}$$
$$x = (x^1, \ldots, x^{k_3})^T \in \mathbf{R}^n,$$

where the m_Ω vectors $y^k = (y_1^k, \ldots, y_{m_\Omega}^k)^T$ and the n_Γ vectors $x^k = (x_1^k, \ldots, x_{n_\Gamma}^k)^T$ consist of the nodal values of $y(u)$ and u at the time level k, respectively, i.e. $y_j^k = y(t^k, N_j; u)$ and $x_j^k = u(t^k, N_j)$.

As before, let

$$\hat{A} = (\hat{A}_{ij})_{i,j=1}^{m_\Omega} = \left(\int_\Omega \nabla \varphi_i \nabla \varphi_j \, d\Omega \right)_{i,j=1}^{m_\Omega}$$

be the stiffness matrix and

$$\hat{M} = (\hat{M}_{ij})_{i,j=1}^{m_\Omega} = \left(\int_\Omega \varphi_i \varphi_j \, d\Omega \right)_{i,j=1}^{m_\Omega}$$

the mass matrix. We use the abbreviations

$$\hat{U}^k = \frac{1}{\Delta t^k} \hat{M} \quad \text{and} \quad \hat{S}^k = \Delta t^k \hat{M}, \quad \text{for } k = 1, \ldots, k_3.$$

Further we define the diagonal matrices

$$\hat{D}^k(x^k) = (\hat{D}_{ii}^k(x^k))_{i=1}^{m_\Omega} = \begin{cases} h_i x_i^k, & i \in I_\Gamma, \\ 0, & \text{otherwise} \end{cases}$$

and $\hat{B} = \nabla_{x^k} D^k(x^k)$, i.e.

$$\hat{B} = (\hat{B}_{ii})_{i=1}^{m_\Omega} = \begin{cases} h_i, & i \in I_\Gamma, \\ 0, & \text{otherwise}, \end{cases}$$

where h_i denotes the average of the neighboring segment lengths at the node N_i. Using these $m_\Omega \times m_\Omega$ matrices as blocks we define the following $m \times m$ block diagonal matrices by

$$M = \begin{bmatrix} 0 & & \\ & \ddots & \\ & 0 & \\ & & \hat{M}^{k_3} \end{bmatrix} \quad U = \begin{bmatrix} \hat{U}^1 & 0 & \\ -\hat{U}^2 & \hat{U}^2 & \\ & \ddots & \ddots \\ & & -\hat{U}^{k_3} & \hat{U}^{k_3} \end{bmatrix} \quad S = \begin{bmatrix} 0 & & & \\ & \ddots & & \\ & & 0 & \\ & & & \hat{S}^{k_2} \\ & & & & \ddots \\ & & & & & \hat{S}^{k_3} \end{bmatrix}$$

and

$$A = \begin{bmatrix} \hat{A} & \\ & \ddots & \\ & & \hat{A} \end{bmatrix} \quad B = \begin{bmatrix} \hat{B} & \\ & \ddots & \\ & & \hat{B} \end{bmatrix} \quad D(x) = \begin{bmatrix} \hat{D}^1(x^1) & & & & \\ & \ddots & & & \\ & & \hat{D}^{k_1}(x^{k_1}) & & \\ & & & 0 & \\ & & & & \ddots \\ & & & & & 0 \end{bmatrix},$$

further we define the m vector

$$R = (\hat{U}^1 H(y_0), 0, \ldots, 0)^{\mathrm{T}}.$$

Then the discrete form of the state problem $(\mathbf{S}(u))$ for a fixed $x \in \mathcal{U}$ reads

$$(\mathcal{S}(x)) \qquad\qquad UH(y(x)) + AK(y(x)) = g(x, y(x)),$$

where

$$g(x, y(x)) = D(x)(y_{wat} - y(x)) + cB(y_{ext}^4 - y(x)^4) + R. \tag{5.15}$$

Because of the block structure of the considerable matrices, the state system can be splitted up to subproblems of time interval t^k, $k = 1, \ldots, k_3$. At every time interval we have a nonlinear subproblem cathering values of the temperature from the previous time level. Because the system $(\mathcal{S}(x))$ is highly nonlinear we use a combination of a modified (SOR) method and a Newton–Raphson type method. The Newton–Raphson method is used when we are calculating the temperature on the boundary. In the interior of the strand we use the modified SOR-method, which can handle the phase changes. For more details we refer to **Zlámal** (1980), **Neittaanmäki and Laitinen** (1988) and **Männikkö** (1991).

The discrete form of the cost function $J(y(u))$ can be written as

$$J(y(x)) = \tfrac{1}{2}(y(x) - y_d)^T B(y(x) - y_d)$$

and the discrete forms of the penalty functions (5.14) by

$$\Psi_1(y(x)) = \left[(y(x) - \Pi_1[y(x)])^T B(y(x) - \Pi_1[y(x)])\right]^{\frac{1}{2}},$$

$$\Psi_2(y(x)) = \left[(\Phi(y(x)) - \Pi_2[\Phi(y(x))])^T B(\Phi(y(x)) - \Pi_2[\Phi(y(x))])\right]^{\frac{1}{2}},$$

$$\Psi_3(y(x)) = \left[(y(x) - \Pi_3[y(x)])^T S(y(x) - \Pi_3[y(x)])\right]^{\frac{1}{2}},$$

$$\Psi_4(y(x)) = \left[(y(x) - \Pi_4[y(x)])^T M(y(x) - \Pi_4[y(x)])\right]^{\frac{1}{2}},$$

(5.16)

where $\Phi(y) \in \mathbf{R}^m$ such that

$$\Phi_i(y) = \frac{y_i - y_{i-1}}{\Delta t^k} \qquad \text{for} \quad i = 1, \ldots, m.$$

Finally let

$$\mathcal{U} = \{x \in \mathbf{R}^n \mid 0 < \alpha \leq x \leq \beta\}$$

be the discrete set of admissible controls. Thus, the matrix form of the problem $(\mathbf{P}_\varepsilon)$ reads in this case

$$(\mathcal{P}_\varepsilon) \qquad \underset{x \in \mathcal{U}}{\text{minimize}} \left\{ J_\varepsilon(y(x)) \equiv J(y(x)) + \sum_{i=1}^{4} \frac{1}{\varepsilon_i} \Psi_i(y(x)) \right\}$$

where $y(x)$ solves the system $(\mathcal{S}(x))$.

5.3.4. *Sensitivity Analysis*

Since the enthalpy function H and the Kirchhoff's transformation K are piecewise linear functions, the state mapping $x \mapsto y(x)$ is nonsmooth. On the other hand, due to the exact penalties, the mapping $y \mapsto E(y) = J_\varepsilon(y)$ is also nonsmooth. Thus we are not able to apply Theorems 2.1 or 2.2 directly. However, in what follows we assume that the discretization is arranged in such a way that these mappings are not nonsmooth at the same points.

Under these assumptions, if x is such that $y \mapsto E(y)$ is smooth (i.e. $y(x)$ lies not on the boundary of any of the constraints sets), then

$$\nabla_y E(y(x)) = \nabla_y J(y(x)) + \sum_{i=1}^{4} \frac{1}{\varepsilon_i} \nabla_y \Psi_i(y(x)) \tag{5.17}$$

where

$$\nabla_y \mathcal{J}(y(x)) = B(y(x) - y_d)$$

and

$$\nabla_y \Psi_1(y(x)) = \frac{1}{\Psi_1(y(x))} B(y(x) - \Pi_1[y(x)]),$$

$$\nabla_y \Psi_2(y(x)) = \frac{1}{\Psi_2(y(x))} B(\Phi(y(x)) - \Pi_2[\Phi(y(x))]),$$

$$\nabla_y \Psi_3(y(x)) = \frac{1}{\Psi_3(y(x))} S(y(x) - \Pi_3[y(x)]),$$ (5.18)

$$\nabla_y \Psi_4(y(x)) = \frac{1}{\Psi_4(y(x))} M(y(x) - \Pi_4[y(x)]).$$

Now, if $\xi_y(x) \in \partial y(x)$, then by Theorem 2.1 we have

$$\xi_y(x)^{\mathrm{T}}[\nabla_y E(y(x))] \in \partial \mathcal{J}_e(y(x)).$$ (5.19)

On the other hand, if x is such that $y(x)$ is on the boundary of some constraint set i, then the corresponding penalty function Ψ_i (and thus also E) is nonsmooth. However, E is regular (convex) and we have $0 \in \partial_y \Psi_i(y(x))$, thus $\xi_E(y(x))$ can be calculated as in (5.17) by choosing $\xi_{\Psi_i}(y(x)) = 0$. Then by Theorem 2.2 we have ($x \mapsto y(x)$ is smooth by assumption)

$$\nabla y(x)^{\mathrm{T}} \xi_E(y(x)) \in \partial \mathcal{J}_e(y(x)).$$ (5.20)

In order to demostrate the adjoint state technique in this case we *suppose for a moment that the mapping* $x \mapsto \mathcal{J}_e(y(x))$ *is smooth*. As before, by differentiating the state system $(\mathcal{S}(x))$ we obtain now

$$U \nabla y(x)^{\mathrm{T}} \nabla_y H(y(x)) + A \nabla y(x)^{\mathrm{T}} \nabla_y K(y(x)) = \nabla_x g(x, y(x)) + \nabla y(x)^{\mathrm{T}} \nabla_y g(x, y(x))$$

which implies that

$$\nabla y(x)^{\mathrm{T}}[U \nabla_y H(y(x)) + A \nabla_y K(y(x)) - \nabla_y g(x, y(x))] = \nabla_x g(x, y(x)).$$ (5.21)

Let $p(x)$ be a solution of the following adjoint state system

$$(\mathcal{A}(x)) \qquad [U \nabla_y H(y(x)) + A \nabla_y K(y(x)) - \nabla_y g(x, y(x))]^{\mathrm{T}} p(x) = \nabla_y E(y(x)),$$

then by (5.19), (5.20) $(\mathcal{A}(x))$ and (5.21) we have

$$\begin{aligned}
\nabla \mathcal{J}_e(y(x)) &= \nabla y(x)^{\mathrm{T}}[\nabla_y E(y(x))] \\
&= \nabla y(x)^{\mathrm{T}}[U \nabla_y H(y(x)) + A \nabla_y K(y(x)) - \nabla_y g(x, y(x))]^{\mathrm{T}} p(x) \\
&= \nabla_x g(x, y(x))^{\mathrm{T}} p(x).
\end{aligned}$$

By differentiating the function g from (5.15) we obtain

$$\nabla \mathcal{J}_e(y(x)) = \nabla D(x)[y_{wat} - y(x)]^T p(x).$$

We are not able to analyse the situation in nonsmooth points, since instead of equalities we obtain inclusions in (5.21) and there is a possibility to obtain a false subgradient if H and K are nonsmooth simultaneously. This is typical in the situations when there exist more nonsmooth points than the dimension of the control variable (see **Haslinger and Roubíček** (1986)). However, in the numerical tests we have observed that this happens very rarely, and therefore, we have been content with this approach.

5.3.5. *Numerical Results*

In our examples we have a billet, whose cross-section is 100×100 mm^2 with rounded corners (the radius of the rounding is 5 mm). The casting speed is 2.8 m/min. The solidus temperature y_{sol} is 1377 °C and the initial temperature y_0 is 1485 °C. We assume that the configuration of the casting machine is symmetric, i.e., we restrict the calculation into the lower left quadrant of the cross-section. The secondary cooling region consists of six zones, whose lengths are 0.25, 0.55, 0.60, 1.15, 1.10 and 1.50 m. Thus, the end of the secondary cooling region is at a distance $z_1 = 5.85$ m (measured from the meniscus level). The unbending point z_3 is 8.50 m and the cutting point z_2 is 10.00 m. The temperature of the secondary cooling water, y_{wat}, is 27°C, and the temperature of the environment, y_{ext}, is assumed to be 97°C in the spray cooling region and 437°C in the radiant cooling region. The minimum values of the heat transfer coefficients are 0.078 kW/m^2°C in zones 1–2, 0.071 kW/m^2°C in zones 3–4 and 0.059 kW/m^2°C in zones 5–6, and the maximum values are 1.445, 1.301 and 1.084 kW/m^2°C, respectively. The maximum length of the liquid pool z_2 is 7.55 m and the minimum temperature y_{duc} at the unbending point is 1025°C.

In Figure 5.3 the constraint functions y_{min} and y_{max} are given, as well as the desired surface temperature y_d on the midfaces and on the corners; between them we have used a parabolic fitting. The constants y'_{min} and y'_{max} (constraints for the cooling and reheating rates on the surface) are -10 °C/s and 10 °C/s, respectively. The dimension of the control x is $n = 25 \times 13 = 325$. (In **Männikkö and Mäkelä** (1990) also some larger problems have been solved.)

The values of the initial cooling (both the heat transfer coefficients and the water flow rates) are presented in Figure 5.4.

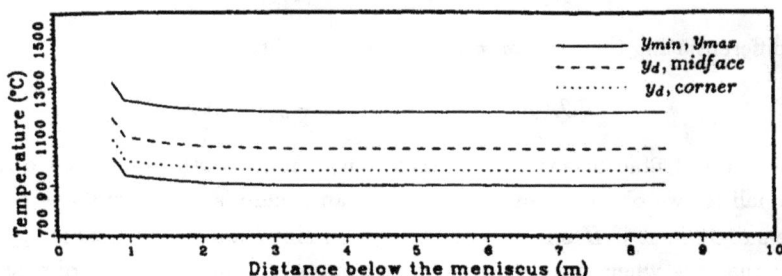

Figure 5.3. Temperatures y_{min}, y_{max} and y_d.

The temperature distribution calculated with this cooling is shown in three points of the strand in the same figure. There appears to be some sharp "peaks" on the surface temperature at the end of different cooling zones, especially after the mold and after the spray cooling region (5.85 m). Furthermore, the shell thickness is also given; as we can see, only after ca. 8.0 m the strand is wholly solid.

The initial cooling has been chosen such that the control constraints are satisfied, but the state constraints are not. The penalty parameters, in turn, are chosen such that all the state constraints are (approximately) equally important at the initial point, i.e. $\varepsilon_1 = 580.0$, $\varepsilon_2 = 1.4 \cdot 10^3$, $\varepsilon_3 = 4.5 \cdot 10^4$ and $\varepsilon_4 = 1.5 \cdot 10^5$.

The test runs with our PB-method have been performed in Cray X-MP EA/432 supercomputer. The CPU-time required for 50 iterations was ca. 80 seconds. The solution after 50 iterations (with exact penalties) is presented in Figure 5.5. For more details we refer to **Männikkö** (1991).

The results of the test run are presented in Table 5.1. The second column of the table gives the values of the penalized cost function during 50 iterations. The iteration '0' means the solution with the initial cooling; the cost function is scaled such that it has a value 1 at that point. In the third column the absolute value of the actual cost function is given. In the remaining columns of the table is the values of the "penalty norms" which tell us how well each of the state constraints is satisfied.

Remark 5.1. *Due to these numerical experiments we may state that Proximal Bundle method has been successfully applied to the control of the secondary cooling in the continuous casting process. Utilizing a supercomputer, it has been possible to use rather dense discretizations and to solve the resulting very large problems in reasonable time.*

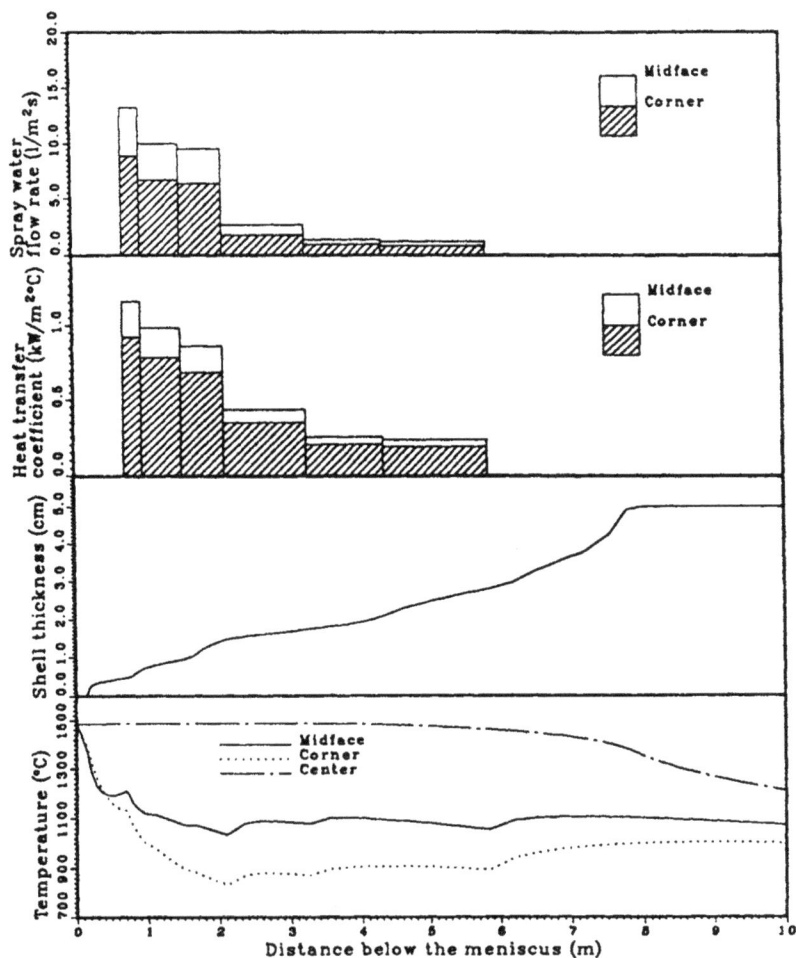

Figure 5.4. Solution with the initial cooling.

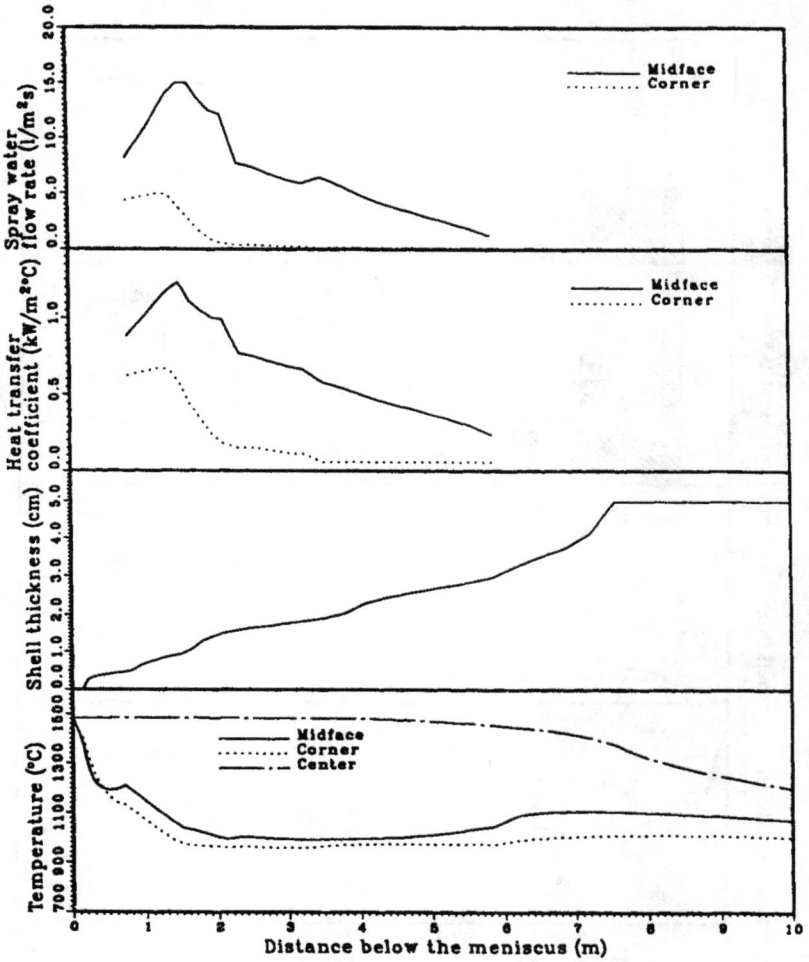

Figure 5.5. Solution after 50 iterations.

Table 5.1.

it	$\mathcal{J}_\epsilon(y(x))$	$\mathcal{J}(y(x))$	$\Psi_1(y(x))$	$\Psi_2(y(x))$	$\Psi_3(y(x))$	$\Psi_4(y(x))$
0	1.0000	12629.	21.741	9.0031	0.28194	0.08502
1	0.68952	10201.	7.7081	7.9548	0.29518	0.03030
2	0.53866	7092.8	1.8397	7.3710	0.10103	0.07375
3	0.46371	7102.9	0.0000	6.8179	0.04304	0.07172
4	0.45642	7667.4	0.0000	6.3722	0.0000	0.08197
5	0.40416	8194.5	0.0000	5.7931	0.01224	0.05814
10	0.35789	10374.	0.0000	3.5265	0.00714	0.04677
15	0.34212	11620.	0.0000	2.2198	0.0000	0.04616
20	0.33875	12185.	0.0000	1.7448	0.00026	0.04532
25	0.33572	12388.	0.0000	1.4802	0.0000	0.04524
30	0.33344	12750.	0.0000	1.0701	0.0000	0.04569
35	0.33281	12823.	0.0000	1.0017	0.00014	0.04553
40	0.33236	12888.	0.0000	0.9331	0.0000	0.04560
45	0.33163	13033.	0.0000	0.7903	0.0000	0.04565
50	0.33113	13107.	0.0000	0.6950	0.00056	0.04566

References

Adams R.A., "Sobolev Spaces," Academic Press, New York, 1975.

Ahmed N.U., "Elements of Finite-Dimensional Systems and Control Theory," Pitman Monographs and Surveys in Pure and Applied Mathematics 37, Longman, Harlow, 1988.

Alekseev V.M., Tikhomirov, V.M. and Fomin, S.V., "Optimal Control," Plenum Press, New York, 1987.

Allen E., Helgason R., Kennington J. and Shetty B., *A Generalization of Polyak's Convergence Result for Subgradient Optimization*, Mathematical Programming **37** (1987), 309–317.

Apostol T. M., "Mathematical Analysis," Addison-Wesley, Massachusetts, 1974.

Arora J., "Introduction to Optimum Design," McGraw-Hill, New York, 1989.

Aubin J.P., "Approximation of Elliptic Boundary Value Problems," J. Wiley & Sons, New York, 1972.

Aubin J.P., "Explicit Methods of Optimization," John Wiley & Sons, Bordas, Paris, 1984.

Aubin J.P. and Ekeland I., "Applied Nonlinear Analysis," John Wiley & Sons, New York, 1984.

Auslender A., *On the Differential Properties of the Support Function of the ε-subdifferential of a Convex Function*, Mathematical Programming **24** (1982), 257–268.

Balinski M. L. and Wolfe P. (Eds.), *Nondifferentiable Optimization*, Mathematical Programming Study **3** (1975).

Banichuk N.V., "Introduction to Optimization of Structures," Springer Verlag, Berlin, 1990.

Barbu V., "Optimal Control of Variational Inequalities," Research Notes in Mathematics 100, Pitman, London, 1984.

Bendsøe M.P. and Kichuchi N., *Generating Optimal Topologies in Structural Design Using a Homogenization Method*, Comput. Methods Appl. Mech. Eng. **71** (1988), 197–224.

Bendsøe M.P. and Rodrigues H.C., *Integrated Topology and Boundary Shape Optimization of 2-D Solids*, Mat. Report No 14 (1989), Danmarks Tekniske Højskola (1989).

Benedict B., Sokolowski J. and Zolésio J.P., *Shape Optimization for Contact Problems*, in "System Modelling and Optimization," (Ed. P. Topft-Chistensen), Lecture Notes in Control and Information Sciences 59, Springer-Verlag, Berlin, 1984, pp. 790–799.

Ben-Tal A. and Zowe J., *Necessary and Sufficient Optimality Conditions for a Class of Nonsmooth Minimization Problems*, Mathematical Programming **24** (1982), 70–91.

Bermudez A. (Ed.), "Optimal Control of Systems Governed by Partial Differential Equations," Proc. of IFIP Workshop on Optimal Control, Lecture Notes in Control and Information Sciences 114, Springer Verlag, Berlin, 1988.

Brandt A.M., "Criteria and Methods of Structural Optimization," Martinus Nijhoff Publishers, Dordrecht, 1986.

Burke J. V., *Second-order Necessary and Sufficient Conditions for Convex Composite NDO*, Mathematical Programming **38** (1987), 287–302.

Burke J., *An Exact Penalization Viewpoint of Constrained Optimization*, SIAM Journal on Control and Optimization **29** (1991), 968–998.

Céa J., *Problems of shape optimal design*, in **Haug and Céa** (1981), Part II, pp. 1005–1048.

Chaney R. W., *Second-order Necessary Conditions in Constrained Semismooth Optimization*, SIAM Journal on Control and Optimization **25** (1987a), 1072–1081.

Chaney R. W., *Second-order Directional Derivatives for Nonsmooth Functions*, Journal of Mathematical Analysis and Applications **128** (1987b), 495–511.

Chaney R. W., *Second-order Necessary Conditions in Semismooth Optimization*, Mathematical Programming **40** (1988), 95–109.

Cheney E. W. and Goldstein A. A., *Newton's Method for Convex Programming and Tchebycheff Approximation*, Numerische Mathematik **1** (1959), 253–268.

Clarke F. H., "Optimization and Nonsmooth Analysis," Wiley-Interscience, New York, 1983.

Clarke F. H., "Methods of Dynamic and Nonsmooth Optimization," CBMS-NSF Regional Conference Series in Applied Mathematics; 57, Philadelphia, Pennsylvania, 1989.

Cominetti R. and Correa R., *A Generalized Second-order Derivative in Nonsmooth Optimization*, SIAM Journal on Control and Optimization **28** (1990), 789–809.

Cornet B., Nguyen V. H. and Vial V. P. (Eds.), *Nonlinear Analysis and Optimization*, Mathematical Programming Study **30** (1987).

Demyanov V. F. and Pallaschke D. (Eds.), "Nondifferentiable Optimization: Motivations and Applications," Springer-Verlag, Berlin, 1984.

Demyanov V. F., Lemaréchal C. and Zowe J., *Approximation to a Set-valued Mapping, I: A Proposal*, Applied Mathematics and Optimization **14** (1986), 203–214.

Demyanov V. F. and Dixon L. C. W. (Eds.), *Quasidifferential Calculus*, Mathematical Programming Study **29** (1986).

Donno F. and Pesamosca G., *On the Solution of the Generalized Steiner Problem by the Subgradient Method*, ZOR – Methods and Models of Operations Research **34** (1990), 335–352.

Duvaut G. and Lions J.L., "Inequalities in Mechanics and Physics," Grundlehren der mathematischen Wissenschaften 219, Springer-Verlag, Berlin, 1976.

Ekeland I. and Temam R., "Convex Analysis and Variational Problems," North Holland, Amsterdam, 1976.

Elliott C.M. and Ockendon J.R., "Weak and Variational Methods for Moving Boundary Problems," Research Notes in Mathematics 59, Pitman, Boston, 1982.

Fiacco A.V., "Introduction to Sensitivity and Stability Analysis in Nonlinear Programming," Academic Press, New York, 1983.

Fletcher R., "Practical Methods of Optimization," John Wiley & Sons, Chichester, 1987.

Fletcher R. and Sainz de la Maza E., *Nonlinear Programming and Nonsmooth Optimization by Successive Linear Programming*, Mathematical Programming **43** (1989), 235–256.

Frankowska H., *The First Order Necessary Conditions for Nonsmooth Variational and Control Problems*, SIAM Journal on Control and Optimization **22** (1984), 1–12.

Friedman A., "Foundations of Modern Analysis," Dover Publications, New York, 1982.

Gaudioso M. and Monaco M. F., *Some Techniques for Finding the Search Direction in Nonsmooth Minimization Problems*, Dip. Sist. Univ. Calabria, Rep. **75** (1988).

Gaudioso M. and Monaco M. F., *Quadratic Approximations in Convex Nondifferentiable Optimization*, SIAM Journal on Control and Optimization **29** (1991), 58–70.

Gill P. E., Murray W. and Wright M. H., "Practical Optimization," Academic Press, London, 1981.

Glowinski R., "Numerical Methods for Nonlinear Variational Problems," Springer-Verlag, New York, 1984.

Ha C. D., *A Generalization of the Proximal Point Algorithm*, SIAM Journal on Control and Optimization **28** (1990), 503–512.

Haftka R.T, Gürdal Z. and Kamat M.P., "Elements of Structural Optimization," Kluwer Academic Publishers, Dordrecht, 1990.

Hartl R. F., *Arrow-type Sufficient Conditions for Nondifferentiable Optimal Control Problems with State Constraints*, Applied Mathematics and Optimization **14** (1986), 229–247.

Haslinger J. and Neittaanmäki P., "Finite Element Approximation of Optimal Shape Design: Theory and Applications," John Wiley & Sons, Chichester, 1988.

Haslinger J., Neittaanmäki P. and Tiba D., *On State Constrained Optimal Shape Design Problems*, in "Proc. Optimal Control of Partial Differential Equations II," (eds. K.H. Hoffmann and W. Krabs), ISNM 78, Birkhauser Verlag, Basel, 1987, pp. 109–122.

Haslinger J. and Roubiček T., *Optimal Control of Variational Inequalities. Approximation Theory and Numerical Realization*, Applied Mathematics and Optimization **14** (1986), 187–201.

Haug E.J. and Céa J. (eds.), "Optimization of Distributed Parameter Structures," Parts I & II. Nato Advances Study Institute Series, Series E, Apl. Sci 50, Sijthoff & Noordhoff, Alphen aan den Rijn, 1981.

Hiriart-Urruty J.-B., *The Approximate First-order and Second-order Directional Derivative for a Convex Function*, in "Mathematical Theories of Optimization," (Eds. Cecconi J. P. and Zolezzi J. P.), Lecture Notes in Mathematics 979, Springer-Verlag, Berlin, 1983, pp. 144–177.

Hiriart-Urruty J.-B., *Approximating a Second-order Directional Derivative for Nonsmooth Convex Functions*, SIAM Journal on Control and Optimization **11** (1984), 43–56.

Hiriart-Urruty J.-B., Strodiot J.-J. and Nguyen V. H., *Generalized Hessian Matrix and Second-order Optimality Conditions for Problems with $C^{1,1}$ Data*, Applied Mathematics and Optimization **11** (1984), 43–56.

Hiriart-Urruty J.-B. and Seeger A., *Calculus Rules on a New Set-valued Second Order Derivative for Convex Functions*, Nonlinear Analysis, Theory, Methods & Applications **13** (1989), 721–738.

Hlaváček I., Bock I. and Lovíšek J., *Optimal Control of a Variational Inequality with Applications to Structural Analysis. Part I. Optimal Design of a Beam with Unilateral Supports*, Applied Mathematics and Optimization **11** (1984), 111–142.

Hlaváček I., Haslinger J., Nečas J. and Lovíšek J., "Numerical Solution of Variational Inequalities," Springer Series in Applied Mathematical Sciences 66, Springer-Verlag, New York, 1988.

Huanwen T., Ye J., Jian G. and Yunjiao H., *Fixed Point Algorithm and Its Applications to Nondifferentiable Programming*, in "Approximation, Optimization and Computing: Theory and Applications," (Eds. Law A. G. and Wang C. L.), Elsevier Science Publishers B. V., North Holland, IMACS, 1990, pp.

291–294.

Ioffe A. D., *New Applications of Nonsmooth Analysis to Nonsmooth Optimization*, in "Mathematical Theories of Optimization,," (Eds. Cecconi J. P. and Zolezzi J. P.), Lecture Notes in Mathematics 979, 1983, pp. 178–201.

Kantorovich L. V., *On the Method of Steepest Descent*, Dokl. Akad. Nauk SSSR **56** (1947), 233–236.

Kaśkosz B. and Lojasiewicz S. Jr., *On a Nonconvex, Nonsmooth Control System*, Journal of Mathematical Analysis and Applications **136** (1988), 39–53.

Kawasaki H., *An Envelope Effect of Infinitely Many Inequality Constraints on Second-order Necessary Conditions for Minimization Problems*, Mathematical Programming **41** (1988), 73–96.

Kelley J. E., *The Cutting Plane Method for Solving Convex Programs*, SIAM J. **8** (1960), 703–712.

Kichucki N. and Oden J.T., "Contact Problems in Elasticity. Study of Variational Inequalities and Finite Element Methods," SIAM, Philadelphia, 1988.

Kim S., Koh S. and Ahn H., *Two-direction Subgradient Method for Non-differentiable Optimization Problems*, Operations Research Letters **6** (1987), 43–46.

Kiwiel K. C., "Methods of Descent for Nondifferentiable Optimization," Lecture Notes in Mathematics 1133, Springer-Verlag, Berlin, 1985.

Kiwiel K. C., *A Method for Solving Certain Quadratic Programming Problems Arising in Nonsmooth Optimization*, IMA Journal of Numerical Analysis **6** (1986a), 137–152.

Kiwiel K. C., *An Aggregate Subgradient Method for Nonsmooth and Nonconvex Minimization*, Journal of Computational and Applied Mathematics **14** (1986b), 391–400.

Kiwiel K. C., *A Method of Linearizations for Linearly Constrained Nonconvex Nonsmooth Minimization*, Mathematical Programming **34** (1986c), 175–187.

Kiwiel K. C., *A Constraint Linearization Method for Nondifferentiable Convex Minimization*, Numerische Mathematik **51** (1987a), 395–414.

Kiwiel K. C., *A Subgradient Selection Method for Minimizing Convex Functions Subject to Linear Constraints*, Computing **39** (1987b), 293–305.

Kiwiel K. C., *A Dual Method for Certain Positive Semidefinite Quadratic Programming Problems*, SIAM Journal on Scientific and Statistical Computing **10** (1989a), 175–186.

Kiwiel K. C., *An Ellipsoid Trust Region Bundle Method for Nonsmooth Convex Minimization*, SIAM Journal on Control and Optimization **27** (1989b), 737–757.

Kiwiel K. C., *Proximity Control in Bundle Methods for Convex Nondifferentiable Minimization*, Mathematical Programming **46** (1990), 105–122.

Kiwiel K. C., *A Tilted Cutting Plane Proximal Bundle Method for Convex Non-differentiable Optimization*, Operations Research Letters 10 (1991), 75–81.

Klatte D. and Tammer K., *On Second-order Sufficient Optimality Conditions for $C^{1,1}$ -optimization Problems*, Optimization 19 (1988), 169–179.

Kohn R.V. and Strang G., *Optimal Design and Relaxation of Variational Problems*, Communications in Pure and Applied Mathematics 39 (1986), Part I 113–137, Part II 139–182, Part III 353–378.

Komlósi S., *On a Possible Generalization of Pshenichnyi's Quasidifferentiability*, Report, Janus Pannonius University, Pécs, Hungary (1989).

Křížek M. and Neittaanmäki P., "Finite Element Approximation of Variational Problems and Applications," Pitman Monographs and Surveys in Pure and Applied Mathematics 50, Longman, Harlow, 1990.

Kurzhanski A., Neumann K. and Pallaschke D. (Eds.), "Optimization, Parallel Processing and Applications," Proceedings of the Oberwolfach Conference on Operations Research, Lecture Notes in Economics and Mathematical Systems 304, Springer-Verlag, Berlin, 1987.

Lemaréchal C., *An Extension of Davidon Methods to Non Differentiable Problems*, in "Nondifferentiable Optimization, Mathematical Programming Study 3," (Eds. Balinski M. L. and Wolfe P.), 1975, pp. 95–109.

Lemaréchal C., *Nondifferentiable Optimization, Subgradient and ε-subgradient Methods*, in "Optimization and Operations Research," Lecture Notes in Economics and Mathematical Systems 117, Springer-Verlag, Berlin, 1976, pp. 191–199.

Lemaréchal C., *Numerical Experiments in Nonsmooth Optimization*, in "Progress in Nondifferentiable Optimization," (Ed. Nurminski E. A.), IIASA-report, 1982.

Lemaréchal C., *Constructing Bundle Methods for Convex Optimization*, in "Fermat Days 85," (Ed. Hiriart-Urruty J. B.), North Holland, Amsterdam, 1986a, pp. 201–240.

Lemaréchal C., *An Introduction to the Theory of Nonsmooth Optimization*, Optimization 17 (1986b), 827–858.

Lemaréchal C., *Mathematical Classification of Optimization Problems*, in "New Methods in Optimization and their Industrial Uses," (Ed. Penot J.-P.), International Series of Numerical Mathematics, Vol. 87, Birkhäuser Verlag, Basel, 1989a, pp. 89–96.

Lemaréchal C., *Nondifferentiable Optimization*, in "Handbooks in OR & MS," (Eds. Nemhauser G. L. et al.), North Holland, 1989b, pp. 529–572.

Lemaréchal C. and Mifflin R. (Eds.), "Nonsmooth Optimization," Proceedings of a IIASA Workshop, Pergamon Press, Oxford, 1977.

Lemaréchal C. and Strodiot J. J., *Bundle Methods, Cutting Plane Algorithms and σ-Newton Directions*, in "Nondifferentiable Optimization: Motivations and Applications," (Eds. Demyanov V. F. and Pallaschke D.), Lecture Notes in Economics and Mathematical Systems 255, Springer-Verlag, Berlin, 1984, pp. 25-33.

Lewis F. L., "Optimal Control," John Wiley & Sons, New York, 1986.

Lions J. L., "Optimal Control of Systems Governed by Partial Differential Equations," Springer-Verlag, New York, 1971.

Lions J. L., "Some Aspects of the Optimal Control of Distributed Parameters Systems," Regional Conference Series in Applied Mathematics, No. 6, SIAM, Philadelphia, 1972.

Lions J.L., "Some Methods in the Mathematical Analysis of Systems and Their Control," Gordon and Breach, New York, 1981.

Lurie K.A., "Applied Optimal Control of Distributed Systems," Plenum Press, New York, 1988.

Mäkelä M. M., "Nonsmooth Optimization: Theory and Algorithms with Applications to Optimal Control," Doctoral Thesis, University of Jyväskylä, Department of Mathematics, Report 47, 1990a.

Mäkelä M. M., *On the Methods of Nonsmooth Optimization*, in "System Modelling and Optimization," (Eds Sebastian H. J. and Tammer K.) Proceedings of the 14th IFIP Conference in Leipzig, Lecture Notes in Control and Information Sciences 143, Springer-Verlag, Berlin, 1990b, pp. 177-186.

Malanovski K., "Stability of Solution to Convex Problems of Optimization," Lecture Notes in Control and Information Sciences 90, Springer-Verlag, Berlin, 1987.

Männikkö T. and Mäkelä M. M., *On the Nonsmooth Optimal Control Problem Connected with the Continuous Casting Process*, in "Advanced Computational Methods in Heat Transfer, Vol. 3: Phase Change and Combustion Simulation," (Eds. Wrobel L. C. et al.) Proceedings of the 1st International Conference in Portsmouth, Computational Mechanics Publications, Southampton & Springer-Verlag, Berlin, 1990, pp. 67-78.

Männikkö T., *Optimal Control of the Continuous Casting Process*, Reports on Applied Mathematics and Computing, University of Jyväskylä, Department of Mathematics 1 (1991).

Meggido N. and Shub M., *Boundary Behavior of Interior Point Algorithms in Linear Programming*, Mathematics of Operations Research 14 (1989), 97-146.

Miettinen K. and Mäkelä M. M., *Nonsmooth Multicriteria Optimization Applied to Optimal Control*, Reports on Applied Mathematics and Computing, University of Jyväskylä, Department of Mathematics 6 (1991).

Mignot F. and Puel J.P., *Optimal Control of Some Variational Inequalities*, SIAM J. Control and Optimiz. **22** (1984), 466–478.

Mifflin R., *An Algorithm for Constrained Optimization with Semismooth Functions*, Mathematics of Operations Research **2** (1977), 191–207.

Mifflin R., *Stationary and Sublinear Convergence of an Algorithm for Univariate Locally Lipschitz Constrained Minimization*, Mathematical Programming **28** (1984), 50–71.

Mifflin R. and Strodiot J. J., *A Bracketing Technique to Ensure Desirable Convergence in Univariate Minimization*, Mathematical Programming **43** (1989), 117–130.

Moreau J. J., Panagiotopoulos P. D. and Strang G. (Eds.), "Topics in Nonsmooth Mechanics," Birkhäuser Verlag, Basel, 1988.

Mota Soares C.A. (ed.), "Computer Aided Optimal Design: Structural and Mechanical Systems," NATO ASI Series F, Vol 27, Springer-Verlag, Berlin, 1987.

Myslinski A., *Bimodal Optimal Design of Vibrating Plates Using Theory and Methods of Nondifferentiable Optimization*, Journal of Optimization and Applications **46** (1985), 187–203.

Myslinski A. and Sokołowski J., *Nondiffrentiable Optimization Problems for Elliptic Systems*, SIAM Journal on Control and Optimization **23** (1985), 632–648.

Nečas J., "Les Méthodes Directes en Théorie des Equations Elliptiques," Masson, Paris, 1967.

Neittaanmäki P., *Computer Aided Optimal Structural Design*, Surveys on Mathematics for Industry **1** (1991), 173–215.

Neittaanmäki P. and Laitinen E., *On Numerical Solution of the Problem Connected with the Control of the Secondary Cooling in the Continuous Casting Process*, Control-Theory and Advanced Technology **4** (1988), 285–305.

Neittaanmäki P. and Stachurski A., *Solving Some Optimal Control Problems Using the Barrier Penalty Function Method*, Applied Mathematics and Optimization **24** (1991).

Neittaanmäki P., Tiba D. and Mäkinen R., *A Variational Inequality Approach to the Problem of the Design of the Optimal Govering of an Obstacle*, in **Bermudez** (1988), 213–224.

Outrata J. V., *On a Class of Nonsmooth Optimal Control Problems*, Applied Mathematics and Optimization **10** (1983), 287–306.

Outrata J. V., *On Using of Bundle Methods in Nondifferentiable Optimal Control Problems*, Problems of Control and Information **15** (1986), 275–286.

Outrata J. V., *On Numerical Solution of Optimal Control Problems with Nonsmooth Objectives: Application to Economic Models*, Kybernetika **23** (1987), 54–66.

Outrata J. V., *A Note on the Usage of Nondifferentiable Exact Penalties in Some Special Optimization Problems*, Kybernetika **24** (1988), 251–258.

Outrata J. V., *On the Numerical Solution of a Class of Stackelberg Problems*, ZOR – Methods and Models of Operations Research **34** (1990), 255–277.

Pallaschke D., Recht P. and Urbański R., *On Locally-Lipschitz Quasi-differentiable Functions in Banach-spaces*, Optimization **17** (1986), 287–295.

Pallaschke D., Recht P. and Urbański R., *On Extension of the Second-order Derivative*, Bulletin of the Polish Academy of Sciences Mathematics **35** (1987), 751–763.

Panagiotopoulos P. D., "Inequality Problems in Mechanics and Applications," Birkhäuser, Boston, 1985.

Panier E. R., *An Active Set Method for Solving Linearly Constrained Nonsmooth Optimization Problems*, Mathematical Programming **37** (1987), 269–292.

Penot J.-P., *On the Mean Value Theorem*, Optimization **19** (1988), 147–156.

Pironneau O., "Optimal Shape Design for Elliptic Systems," Springer Series in Computational Physics, Springer-Verlag, New York, 1984.

Polak E., *On the Mathematical Foundations of Nondifferentiable Optimization in Engineering Design*, SIAM Review **29** (1987), 21–89.

Polyak R. A., *Smooth Optimization Methods for Minimax Problems*, SIAM Journal on Control and Optimization **26** (1988), 1274–1286.

Raĭtum V. E., *On Optimal Control Problems for Linear Elliptic Equations*, Soviet Math. Dohl. **20** (1979), 129–132.

Roberts A. W. and Varberg D. E., "Convex Functions," Academic Press, New York, 1973.

Robinson S. M., *Bundle-based Decomposition: Conditions for Convergence*, Working paper in IIASA, Laxenburg, Austria (1987).

Rockafellar R. T., "Convex Analysis," Princeton University Press, Princeton, New Yersey, 1970.

Rockafellar R. T., "The Theory of Subgradients and its Applications to Problems of Optimization. Convex and Nonconvex Functions," Heldermann Verlag, Berlin, 1981.

Rockafellar R. T., *Generalized Subgradients in Mathematical Programming*, in "Mathematical Programming, The State of Art," (Eds. Bachem A., Grötschel M. and Korte B.), Springer-Verlag, Berlin, 1983, pp. 368–390.

Rockafellar R. T., *Maximal Monotone Relations and the Second Derivatives of Nonsmooth Functions*, Annale Institute Henri Poincare **2** (1985), 167–184.

Rockafellar R. T., *Optimization: A Case for the Development of New Mathematical Concepts*, Journal of Computational and Applied Mathematics **22** (1988), 243–255.

Rockafellar R. T., *Second-order Optimality Conditions in Nonlinear Programming Obtained by Way of Epi-derivatives*, Mathematics of Operations Research **14** (1989), 462–484.

Rodrigues J.F., "Obstacle Problems in Mathematical Physics," North Holland Mathematical Studies 134, North Holland, Amsterdam, 1987.

Roubiček T., *Evaluation of Clarke's Generalized Gradient in Optimization of Variational Inequalities*, Kybernetika **25** (1989), 157–168.

Sachs E., *Global Convergence of Quasi-Newton-type Algorithms for Some Nonsmooth Optimization Problems*, Journal of Optimization Theory and Applications **40** (1983), 201–219.

Schirotzek W., *Nonasymptotic Necessary Conditions for Nonsmooth Infinite Optimization Problems*, Journal of Mathematical Analysis and Applications **118** (1986), 535–546.

Schramm H., "Eine Kombination von Bundle- und Trust-Region-Verfahren zur Lösung nichtdifferenzierbarer Optimierungsprobleme," Bayreuther Mathematische Schriften, Heft 30, Bayreuth, 1989.

Schramm H. and Zowe J., "A Version of the Bundle Idea for Minimizing a Nonsmooth Function: Conceptual Idea, Convergence Analysis, Numerical Results," DFG-report, Universität Bayreuth, 206, 1990.

Shor N. Z., *Generalized Gradients Methods of Nondifferentiable Optimization Employing Space Dilatation Operations*, in "Mathematical Programming, The State of Art," (Eds. Bachem A., Grötschel M. and Korte B.), Springer-Verlag, Berlin, 1983, pp. 368–390.

Shor N. Z., "Minimization Methods for Non-differentiable Functions," Springer-Verlag, Berlin, 1985.

Sokolowski J., *Differential Stability of Solutions to Constrained Optimization Problems*, Applied Mathematics and Optimization **13** (1985), 97–115.

Sokolowski J., *Sensitivity Analysis of Contact Problems with Prescribed Friction*, Applied Mathematics and Optimization **18** (1988), 99–117.

Studniarski M., *Mean Value Theorems and Sufficient Optimality Conditions for Nonsmooth Functions*, Mathematical Analysis and Applications **111** (1985a), 313–326.

Studniarski M., *Mean Value Theorem for Functions Possessing First Order Convex Approximations. Applications in Optimization Theory*, Zeitschrift für Analysis und ihre Anwendungen 4 (1985b), 125–132.

Studniarski M., *Sufficient Conditions for the Stability of Local Minimum Points in Nonsmooth Optimization*, Optimization **20** (1989a), 27–35.

Studniarski M., *An Algorithm for Calculating One Subgradient of a Convex Function of Two Variables*, Numerische Mathematik **55** (1989b), 685–693.

Tarasov V. N. and Popova N. K., *A Modification of the Cutting-plane Method with Accelerated Convergence*, in "Nondifferentiable Optimization: Motivations and Applications," (Eds. Demyanov V. F. and Pallaschke D.), Springer-Verlag, Berlin, 1984, pp. 284–290.

Tiba D., "Optimal Control of Nonsmooth Distributed Parameter Systems," Lecture Notes in Mathematics, 1325, Springer Verlag, 1990.

Teo K. L. and Goh C. J., *On Constrained Optimization Problems with Nonsmooth Cost Functionals*, Applied Mathematics and Optimization **18** (1988), 181–190.

Teo K.L. and Wu Z.S., "Computational Methods for Optimizing Distributed Systems," Academic Press, New York, 1984.

Tröltzsch F., "Optimality Conditions for Parabolic Control Problems and Applications," Teubner-Texte zur Matematik, 62, Leipzig, 1984.

Uryas'ev S. P., *New Variable Metric Algorithms for Nondifferentiable Optimization Problems*, Journal of Optimization Theory and Applications (1991) (to appear).

Ward D. E. and Borwein J. M., *Nonsmooth Calculus in Finite Dimensions*, SIAM Journal on Control and Optimization **25** (1987), 1316–1340.

Ward D. E., *Which Subgradients Have Sum Formulas?*, Nonlinear Analysis, Theory, Methods & Applications **12** (1988), 1231–1243.

Warga J., *Second Order Necessary Conditions in Optimization*, SIAM Journal on Control and Optimization **22** (1984), 524–528.

Wolfe P., *A Method of Conjugate Subgradients for Minimizing Nondifferentiable Functions*, in "Nondifferentiable Optimization, Mathematical Programming Study 3," (Eds. Balinski M. L. and Wolfe P.), 1975, pp. 145–173.

Yuan Y., *On the Superlinear Convergence of a Trust Region Algorithm for Nonsmooth Optimization*, Mathematical Programming **31** (1985), 269–285.

Yuan Y., *Some Results in Nonsmooth Optimization*, Journal of Computational Mathematics **5** (1987), 74–88.

Zagrodny D., *A Note on the Equivalence between the Mean Value Theorem for the Dini Derivative and the Clarke-Rockafellar Derivative*, Optimization **21** (1990), 179–183.

Zlámal M., *A Finite Element Solution of the Nonlinear Heat Equation*, R.A.I.R.O. Analyse numerique/Numerical Analysis **14** (1980), 203–216.

Zowe J., *Nondifferentiable Optimization*, in "Computational Mathematical Programming," 1985, pp. 323–356.

Zowe J., *Optimization with Nonsmooth Data*, OR Spektrum **9** (1987), 195–201.

Zowe J., *The BT-algorithm for Minimizing a Nonsmooth Functional Subject to Linear Constraints*, Working paper in Department of Economics, University of Bergen, Norway **1088** (1988).

Index